Botond Lajos Borcsa

Diterpene alkaloids of Aconitum anthora L. and A. moldavicum L.

Botond Lajos Borcsa

Diterpene alkaloids of Aconitum anthora L. and A. moldavicum L.

Phytochemical studies and certain pharmacological investigations carried out on selected diterpene and lipo-alkaloids

LAP LAMBERT Academic Publishing

Impressum / Imprint

Bibliografische Information der Deutschen Nationalbibliothek: Die Deutsche Nationalbibliothek verzeichnet diese Publikation in der Deutschen Nationalbibliografie; detaillierte bibliografische Daten sind im Internet über http://dnb.d-nb.de abrufbar.
Alle in diesem Buch genannten Marken und Produktnamen unterliegen warenzeichen-, marken- oder patentrechtlichem Schutz bzw. sind Warenzeichen oder eingetragene Warenzeichen der jeweiligen Inhaber. Die Wiedergabe von Marken, Produktnamen, Gebrauchsnamen, Handelsnamen, Warenbezeichnungen u.s.w. in diesem Werk berechtigt auch ohne besondere Kennzeichnung nicht zu der Annahme, dass solche Namen im Sinne der Warenzeichen- und Markenschutzgesetzgebung als frei zu betrachten wären und daher von jedermann benutzt werden dürften.

Bibliographic information published by the Deutsche Nationalbibliothek: The Deutsche Nationalbibliothek lists this publication in the Deutsche Nationalbibliografie; detailed bibliographic data are available in the Internet at http://dnb.d-nb.de.
Any brand names and product names mentioned in this book are subject to trademark, brand or patent protection and are trademarks or registered trademarks of their respective holders. The use of brand names, product names, common names, trade names, product descriptions etc. even without a particular marking in this work is in no way to be construed to mean that such names may be regarded as unrestricted in respect of trademark and brand protection legislation and could thus be used by anyone.

Coverbild / Cover image: www.ingimage.com

Verlag / Publisher:
LAP LAMBERT Academic Publishing
ist ein Imprint der / is a trademark of
OmniScriptum GmbH & Co. KG
Heinrich-Böcking-Str. 6-8, 66121 Saarbrücken, Deutschland / Germany
Email: info@lap-publishing.com

Herstellung: siehe letzte Seite /
Printed at: see last page
ISBN: 978-3-659-63345-4

Zugl. / Approved by: Szeged, University of Szeged, Diss., 2014

TABLE OF CONTENTS

ABBREVIATIONS AND SYMBOLS

1D, 2D	one-dimensional, two-dimensional
Alkaloid types:	DAs = diterpene alkaloids; TDAs, DDAs, MDAs, LAs = triester, diester, monoester, lipo-type diterpene alkaloids
COSY	correlated spectroscopy
Ester groups:	Ac = acetyl, As = anisoyl, Bz = benzoyl, Cn = cinnamoyl, Vr = veratroyl
Functionalities:	Et = ethyl, Me = methyl, OMe = methoxy
hERG	human ether-à-go-go-related gene
HMBC	heteronuclear multiple-bond correlation
HREIMS	high-resolution electron ionisation mass spectrometry
HRMS	high-resolution mass spectrometry
HSQC	heteronuclear single-quantum correlation
J	coupling constant
JMOD	*J*-modulated spin-echo experiment
m.p.	melting point
Methods:	CC = open-column chromatography, CPC = centrifugal planar chromatography, GFC = gel-filtration chromatography, PLC = preparative layer chromatography, TLC = thin-layer chromatography, VLC = vacuum liquid chromatography, NP = normal phase, RP = reversed phase
nAChR	nicotinic acetylcholine receptor
$Na_v1.2$	voltage-gated sodium channel isoform 1.2
NMR	nuclear magnetic resonance
NOESY	nuclear *Overhauser* effect spectroscopy
QSAR	quantitative structure-activity relationship
Solvents:	EtOH = ethanol, EtOAc = ethyl acetate; MeOH = methanol, *n*Hex = normal hexane, To = toluene, CH_2Cl_2 = dichloromethane, $CHCl_3$ = chloroform
syn.	synonym
TMS	tetramethylsilane
UV	ultraviolet
δ	chemical shift

3

1. INTRODUCTION

The buttercup family (*Ranunculaceae*) has more than 2000 species and by producing specific compounds possessing a very wide range of pharmacological characteristics and physiological effects[1†], thus it has been in the focus of an ever-growing scientific interest worldwide. Species belonging to the most actively investigated genera (*Aconitum, Delphinium, Consolida*) are widely distributed throughout the Northern hemisphere. Based on common empiric knowledge these plants are considered to be poisonous. Species of these genera accumulate highly toxic DAs, which have attracted considerable interest motivated by their complex structures, interesting chemistry and noteworthy physiological effects. This exponentially increasing interest is well illustrated by the fact that since 1833, when GEIGER had discovered the archetypical representative of these compounds, aconitine, more than 900 natural compounds were discovered by 2008.[2]

Although being violently toxic in character, a widely distributed species, *A. napellus*, gained certain healing reputation at the beginning of the European professional official therapy as its roots (*Aconiti tuber*) and main alkaloid, aconitine, were used for the treatment of trigeminal neuralgia.[3] From the beginning to the late second third of the 20th century, in parallel with the recognition of the very narrow therapeutic range of DAs, the medicinal use of aconite drugs had disappeared from Western medicinal practices. Nevertheless, in Far Eastern traditional medicinal systems (Traditional Chinese Medicine (TCM), Ayurveda, Kampo, Unani etc.) aconite drugs have been applied for centuries as painkillers and antirheumatic agents.[4] Several unprocessed and processed aconite drugs are still official in the national pharmacopoeias of numerous Far Eastern countries.

In spite of our rapidly growing knowledge concerning the chemistry of processed drugs there is still no sufficient information on their pharmacology and toxicology. With the worldwide increasing spread of these medicinal systems, and with the increasing public interest towards phytomedicines, phytoanalytical studies on aconite drugs are of growing importance with respect to their safe application. In consequence of the harmonization of TCM and modern, evidence based medicinal systems aconite preparations are attracting increasing interest today in Western medicine too. The European Directorate for the Quality of Medicines and Health Care (EDQM) devotes special attention to the quality control of herbal drugs of TCM, within the frame of which the incorporation of *Aconitum* drugs into the European Pharmacopoeia is currently in progress.

Nowadays special emphasis has been placed to research activities aiming the study of the effects of traditional processing methods on the chemical composition of aconite drugs, elaboration of adequate analytical methods for the quality control of these processing methods, and pharmacological-toxicological evaluations of their alkaloids.

2. AIMS OF THE STUDY

In 2001 the research group of HOHMANN *et al.* (Department of Pharmacognosy, University of Szeged) has started a research program dealing with *Aconitum* species native to the Carpathian Basin. The aim of this programme is to study these species since chemical and pharmacological characteristics of the rare and/or endemic *Aconitum* species are partially or completely unexplored. Within the frame of this programme CSUPOR carried out a comprehensive screening of 6 taxa from which *A. toxicum* and *A. vulparia* were the first selected species to be thoroughly examined as part of his doctoral thesis.[3] By joining to this ongoing comprehensive research the main goals of my work were to

1. provide review of recent advances in the topic of diterpene alkaloids;
2. phytochemically examine the alkaloid contents of *A. anthora* and *A. moldavicum*, i.e. isolate their DAs via preparative chromatographic methods in order to gain information concerning the chemistry of the species of the Ranunculaceae family;
3. elucidate the structures of isolated compounds via NMR and HRMS techniques, provide characteristic spectral data on the isolated new compounds, and supplement missing NMR data on the already-known compounds;
4. carry out the semisynthesis and purification of a series of LAs derived from aconitine;
5. in the frame of co-operations, carry out and evaluate *in vitro* anti-inflammatory activity of LAs and hERG and $Na_v1.2$ channel activities of previously or newly isolated DAs;
6. carry out quantitative determination of toxic DDAs in authentic processed *A. carmichaelii* and *A. kusnezoffii* samples.

3. LITERATURE OVERVIEW[*]

3.1. Botany of the *Aconitum* genus and the investigated species

3.1.1 Supra-generic taxonomic classification of the genus Aconitum

Division: Spermatophyta, subdivision: Angiospermatophyta, class: Dicotyledonopsida, subclass: Ranunculidae, superorder: Ranunculanae, order: Ranunculales, family: Ranunculaceae.[5]

Recently the subfamilial rank of the tribe of Delphinieae is used to comprise the species-rich genera, *Aconitum* and *Delphinium* (the latter including *Consolida* and *Aconitella*). By comprising 650-700 species this tribe represents to some 25% of all Ranunculaceae species.[6] Species belonging to this tribe are mostly holarctic in distribution and occupying mainly circumboreal and/or alpine habitats, with only a few species to occur on mountains from tropical Africa. The center of species diversity and diversification is deducted to Southwest China and eastern Himalayas, where *Aconitum* has 166 species (from all the approximated 300-350 taxa).[6] Life histories of Delphinieae species show relatively high diversity varying from short-living strictly annual, pseudo-annual and facultative annual forms occurring mainly in the Mediterranean and the xeric Irano-Turanian regions; to biennial and perennial forms of longer life-span (even up to six-seven decades) occurring mostly in cold and wet high altitude habitats in Southeast Asia and North America.

3.1.2 Infra-generic taxonomic considerations

Considering the genus *Aconitum* a widely accepted classification specifies three infra-generic ranks: the subgeneras of the biennial/pseudo-annual tuberous root growing *Aconitum*, the perennial rhizome having *Lycoctonum* and the monospecific facultative annual or biannual *Gymnaconitum*.[3,6,7] According to the recent results on the phylogeny of Delphinieae obtained by using molecular phylogenetic and phylogeographic methods (besides considering the extremely variable morphological aspects) the tribe is proposed to consist of three natural genera: *Staphisagria*, *Aconitum* and *Delphinium*.[6]

According to the taxonomic system employed in the latest edition of Flora Europaea[8] only 7 taxa are recognised as individual species of the genus *Aconitum*, 5 of which is native to the Carpathian Basin.[3] In this system the case of *A. anthora* (subgenus: *Aconitum*) is fairly uncomplicated which is supported by the plant's distinct appearance and morphology, while the case of *A. moldavicum* (subgenus:

---

[*] In his thesis, CSUPOR summarised the topic until the year 2007, therefore the present work aims to review the literature published after this date. Some recitation could not be avoided, these are marked with a [†].

Lycoctonum) may be regarded as more problematic, because it is referred as a subspecies in the form of *A. lycoctonum* L. subsp. *moldavicum* (Hacq.) Jalas. UTELLI *et al.* conducted a molecular systematics study on all European and Caucasian species belonging of the subgenus *Lycoctonum*. They found that in spite of the obvious morphological difference in flower colour between *A. lycoctonum* (yellow) and *A. moldavicum* (blue), these taxa are genetically indistinguishable. Since flower colour is an evolutionarily labile, thus variable character in *Aconitum* and because of the smaller distribution area of *A. moldavicum*, which is situated at the eastern border of the distribution of *A. lycoctonum*, they suggested that *A. moldavicum* may be a derived form of *A. lycoctonum* as a quantum speciation product appeared during the ice ages.[7] From this synonymy we have chosen and in the present work using the form *A. moldavicum* Hacq. because it is a more widely accepted form in Hungary.

3.1.3 Description of the studied species

Aconitum anthora L. (anthora, yellow monkshood, healing wolfsbane) is a 50 – 120 cm tall, upstanding, normally not branching, soft stemmed plant. Its tuberaceous roots are small; leaves are deeply palmately lobed into many thin filaments. The basal leaves and the radical leaves on the stem have 2 or 3 times longer petioles than the leaf blades, and they are generally dry before flowering. The upper leaves on the stem have a short petiole (10 mm), and they are progressively become smaller towards to the top of the stem. Compact, terminal, simple or branched off racemes contain pale yellow flowers, which have typical dorsiventral and spirocyclic symmetry, and they are also peculiar to a helmet. Atrophied petals inside the helmet are forming nectaries. The spur of the two petals is contorted. Its fruit is follicle. The plant has a blooming period between the end of August to the beginning of October. *A. anthora* occurs in calciferous xerotherm forests and lawns, loose and arid rupestral, rubble, loam, clay and forest soils. It is an alpine plant living in the southern part of Central Europe. It is distributed from the Pyrenees to the Far East.[7,9,10]

A. moldavicum Hacq. (Carpathian or eastern monkshood) is 60 – 150 cm tall, upstanding, normally not branching, soft stemmed plant. Its rhizomatous roots are fleshy; its leaves are dark green and palmately lobed into their half or two-third, and having 4-6 bigger segments. The inflorescence is covered with short, crispate eglandular hairs, and it appears as terminal, simple or branched off racemes containing characteristic flowers, which have typical dorsiventral and spirocyclic symmetry, and they are also peculiar to a helmet. The colour of the flowers may vary from reddish to deep bluish purple. Atrophied petals inside the helmet are forming nectaries. The spur of the two petals is contorted. It has also follicular fruits. The plant has a blooming period between the beginning of June to the end of July.

A. moldavicum occurs mainly in montane beeches. It is a montane-subalpine (subendemic) Carpathian plant living in the eastern and central parts of Europe with a habitat extending to Romania and West Ukraine.[7,9,10]

A. *anthora* and A. *moldavicum* are protected species in Hungary. Our examinations were carried out by the permission of the environmental protection authorities.

3.2 Recent advances in the chemistry of the *Aconitum* genus and the investigated species

Ranunculaceae species contain alkaloids which can be divided into three main classes: C_{18} bisnorditerpene, C_{19} norditerpene and C_{20} DAs. Further grouping of compounds belonging to each main class is based on the type of the carbon skeletons (for examples see **Annex I**).[18] DAs are aminated derivatives of nitrogen-free terpenes, which means that these compounds should be considered biogenetically as pseudoalkaloids. It is very interesting that some diterpenoid alkaloids have also been isolated from greenflies (*Brachycaudus aconiti* and B. *napelli*, Aphididae) living on *Aconitum* species.[11]

3.2.1 Diterpene alkaloids

A) C_{18} bisnorditerpene alkaloids
To date these compounds are represented by two skeletal types.
1. Lappaconitine-type compounds (I) which are structurally characterised by the presence of a C_1 unit at C-4. The majority of lappaconitines are reported from *Aconitum* and *Delphinium* species. Acotoxicine (**1**) isolated by CSUPOR *et al.* from A. *toxicum*[12] and piepunendine B (**2**) containing an exceptional 2-(p-hydroxyphenyl)ethoxy substituent at C-8 can be mentioned in this group. Kiridine (**3**) is the first example of DAs that contain a 9,14-methylenedioxy group besides possessing a C-3,C-4 epoxide functionality.[18]

2. Ranaconitine-type compounds (II) have an oxygen-containing functionality at C-7. An interesting example is linearilin (**4**) containing the rare hydroperoxyl group at C-7.[13,18]

B) C_{19} norditerpene alkaloids

1. Aconitine-type compounds (III) are the most common representatives of the C_{19} NDAs with the aconitane core. Among all DAs known today more than 50% contain the aconitane core. 115 new representatives belonging to this skeletal type were reported only within the period from 1998 to 2008.[18] The first diterpenoid alkaloid, aconitine (**5**) was isolated from *A. napellus* in 1821 by PESCHIER, but its structure was discovered only later. Despite of numerous chemical modifications, it was only as early as 1959, when the X-ray analysis of a derivative, demethanolaconinone-hydroiodide-trihydrate helped to define the entire structure, including the configuration of the 13 chirality centres.[14] These compounds have no oxygen functionality at C-7, most of them are characterised by an α-oriented hydroxyl group at C-6. Based on the nitrogen functionality further four subtypes have been described: amines, *N,O*-mixed acetals, imines and amides.[18]

2. Lycoctonine-type compounds (IV) embody probably the second largest group of DAs, which are characterised by the presence of an oxygenated group at C-7. These can also be subdivided to amine, *N,O*-mixed acetal, imine and amide subtypes. Further distinctions are based on the oxygen-containing groups to be found at C-7/C-8. 7,8-Methylenedioxy moiety containing compounds were mainly reported from *Delphinium* species, but isodelatine (**6**) was isolated from the roots of *A. taipaicum*, presenting a rare occurrence of such an alkaloid in an *Aconitum* species. Jiufengsine (**7**) is the first example of lycoctonine-type alkaloids to contain an anthranoyl group at C-8.[18]

3. Pyro-type compounds (V) contain a $\Delta^{8,15}$ double bond or an 8-H/15-ketone unit and are considered to be decomposition derivatives of aconitine-type compounds which have lost their C-8 substituent. These compounds were only reported from processed aconite drugs, therefore can be recoginsed as artefacts.[18]

4. Lactone-type compounds (VI) are very rare alkaloids having a δ-lactonised C ring. Until 2008, only nine such compounds were reported.[18]

5. *7,17-seco-type compounds* (VII) have a $\Delta^{7,8}$ double bond and biosynthetically originate from aconitine-type alkaloids via Grob fragmentation. Representatives of this group are secokaraconitine **(8)** and secoyunaconitine **(9)**, the former possessesing the unusual *C-17=N* unit.[18]

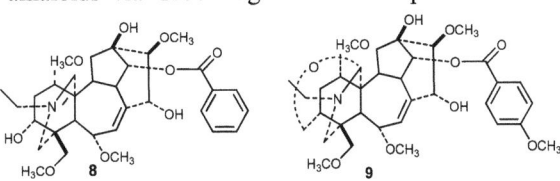

6. *Rearranged-type compounds* (VIII) contain unusual bridges (e.g. C-8–C-17 instead of C-7–C-17. A recently reported compound, vilmoraconitine **(10)** found in *A. vilmorinianum* is the first DA to contain a three-membered ring (C-8, C-9, C-10).[18]

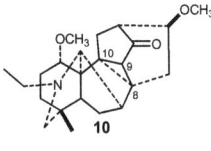

C) C_{20} diterpene alkaloids

Alkaloids belonging to this chemical class are usually less oxidised having only 2-5 oxygen substituents, but by formation of more variant core types these compounds may also claim significant perspectives. Major subclasses are the followings:

1. *Atisine-type compounds* (IX) have a pentacyclic core and form one of the most common DA group to occur outside the Ranunculaceae family in *Spiraea* species (Rosaceae). An unprecedented structure was proven for delphatisine B **(11)** having a γ-lactone-fused oxazolidine ring.[18]

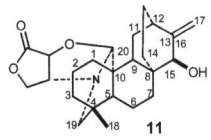

2. *Denudatine-type compounds* (X) are a subclass of hexacyclic C_{20} DAs based on atisines with an additional bond between C-20 and C-7.[18]

3. *Hetidine-type compounds* (XI) are also hexacyclic C_{20} DAs based on atisines having an additional bond between C-20 and C-14.[18]

4. *Hetisine-type compounds* (XII) are heptacyclic compounds formed by an extra bond between the *N* and C-6 atoms and represent one of the most complex structural group derived from atisine core. An example of these compounds, guan-fu base S **(12)** was obtained along with a new *ent*-kaurane diterpene, which may support the hypothesised biogenetic pathway of C_{20} DAs involving nitrogen-free diterpenes.[18]

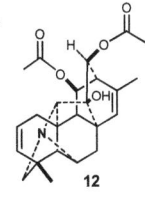

5. *Vakognavine-type compounds* (XIII) have an *N*,19-seco hetisine skeleton in addition to a C-4 aldehyde group.[18]

6. *Napelline-type compounds* (XIV) in comparison with veatchine-type compounds have a hexacyclic skeleton with an additional C-20–C-7 bridge.[18]

7. Kusnezoline-type compounds (**XV**) having an adamantane skeleton were firstly prepared from hetisine via acid-catalysed rearrangement but later found in nature, as it is illustrated by the isolation of guan-fu base K (**13**).[18]

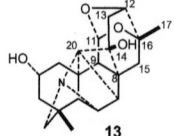

8. Racemulosine-type compounds (**XVI**) are novel C_{20} DAs with a unique skeleton and proposed to be originated from the denudatine-type DAs through double Wagner-Meerwein rearrangements of rings A and C. So far racemulosine (**14**) reported from *A. racemulosum* var. *pengzhouense* is the only member of this subclass.[18]

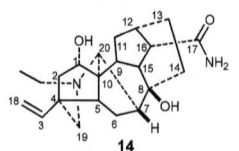

9. Arcutine-type compounds (**XVII**) also represent unusual structures having a C-5–C-20 bond in lieu of the traditional C-10–C-20 bond being typical to C_{20} DAs.[18]

A part of the C_{20} diterpenoid alkaloids, such as the atidane (**XVIII**) [including ajaconine (**XIX**)], hetisane (**XX**), vilmoridine (**XXI**), and delnudine (**XXII**) types are probably structurally related to atisane diterpenoids. However, compounds with a veatchine (**XXIII**), and 14,20-cycloveatchine (**XXIV**) skeleton show structural relationship with the kaurane diterpenoids.[15]

Based on the number of ester groups present in the molecules, DAs can be classified otherwise into the main types of diester (DDAs), monoester (MDAs), and triester DAs (TDAs). As concerns the latter group only a few compounds were detected by highly sensitive analytical methods, but no such compound was isolated.

It is noteworthy to mention that some very intriguing bisditerpenoid alkaloids have also been reported, which are biosynthesised by connecting 2 C_{20} DA units or of 1 C_{20} and 1 C_{19} unit.[16]

3.2.2 Lipo-type diterpene alkaloids

In 1982, when examining *Aconiti tuber* (Chuanwu) KITAGAWA *et al.* isolated alkaloid mixtures containing long-chain fatty acid residues. Based on the results of the structure elucidation of isolated components carried out via certain indirect methods of chemical degradation the authors proposed the general term 'lipo-alkaloid' to describe the isolates (lipoaconitine, lipohypaconitine, lipomesaconitine, lipodeoxyaconitine).[17] Scientific data and knowledge on these compounds has shown continuously intensifying accumulation since the time of their first recognition, therefore it was interesting to see that although in their recently published review covering the period from 1998 to 2008, WANG *et al.*[18] summarised comprehensively the current information on the native C_{18}, C_{19} and C_{20} diterpenoid alkaloids; yet, LAs were referred only shortly in their work. Therefore, our research group has compelled a review paper to summarise the structures, chemistry, semisynthesis, analytics and

12

bioactivities of LAs revealed to date, thus based on 32 references we provided the first comprehensive study on this topic covering the data of 173 compounds.[19]

From the chemical-structural point of view LAs are transesterified derivatives of C_{19} aconitane DDAs containing one or two long-chain fatty acid residues. All LAs identified so far are based on the C_{19} aconitane skeleton, on which the positions C-1, C-3, C-10, C-13, C-14, C-15 and the N atom can be substituted by different functional groups. The fatty acid residues primarily bond at C-8, but according to the recent report of XU et al.[20] these residues can also connect at position C-3, forming dilipo derivatives. The already reported 173 LAs are formed by the combination of 12 parent DDAs and 36 fatty acids (compounds produced only semisynthetically are also taken into account). These fatty acids vary from C_3 propanoic acid to C_{25} pentacosanoic acid, the grade of unsaturation also varies in a broad spectrum from the compounds being monounsaturated up to compounds containing 6 double bonds. Structures of the already known LAs along with their parent DDAs and the structures of by-products of semisynthetic reactions were also summarized in our before mentioned paper.[19]

Many papers on the isolation of LAs from plant material[21-25] reported these compounds to be complex mixtures, and stated that direct chromatographic isolation of the individual components is extremely difficult because of their very similar physico-chemical properties and chromatographic behaviour. Therefore, at the beginning of the research on the topic indirect degradation (methanolic[17], alkaline[24,25]) methods were elaborated, while later on and along with the development of highly sensitive analytical methods electrospray ionization mass spectrometry (ESI-MS^n)[20,22,26-28], matrix assisted laser desorption/ionization mass spectrometry (MALDI-MS^n)[23,29,30], and atmospheric-pressure chemical ionization mass spectrometry (APCI-MS^n)[31] were and are the main methods used for unambiguous compound identification and structure elucidation. However, as a major drawback of such mass spectrometric methods it should be noted that mass spectrometric data are not appropriate to obtain exact details about the positions of double bonds in the esterifying fatty acid side chains (determination of bonding isomers of e.g. linolenic acid). This aspect may have special significance since it could have major influence on the pharmacological effects of the compounds concerned, as it can be observed in the case of certain n-3 and n-6 fatty acids.[32]

As for origin, LAs are native minor compounds of crude aconite drugs (A. carmichaelii - roots[22,30,31,33], A. ferox - roots[21], A. kusnezoffii - flowers[28]), yet as part of the study of the traditional Chinese drug processing methods of aconite roots, WANG et al. proved for the first time that during the decoction of the roots artificial production of LAs can be observed.[33] This was demonstrated by adding palmitic acid to aconite root before decoction, and the resulting processed drug was found to be containing 14-benzoylaconine-8-*O*-palmitate, 14-benzoylhypaconine-8-*O*-palmitate, and 14-benzoyldeoxyaconine-8-*O*-palmitate in amounts which were not present previously in the alkaloid mixture of the processed drugs. This experiment served as direct evidence for the previous speculations that LAs can be formed by transesterification reactions during processing of the crude drugs. The facts that LAs can be found in both unprocessed and processed aconite drugs, and that processing results in the increase of LA content in parallel with the decrease of DDAs concentration besides the formation of hydrolysed products also were confirmed by the work of Csupor *et al.*[31] **Figure 1** shows the general scheme of the transesterification reaction using the example of aconitine (**5**).

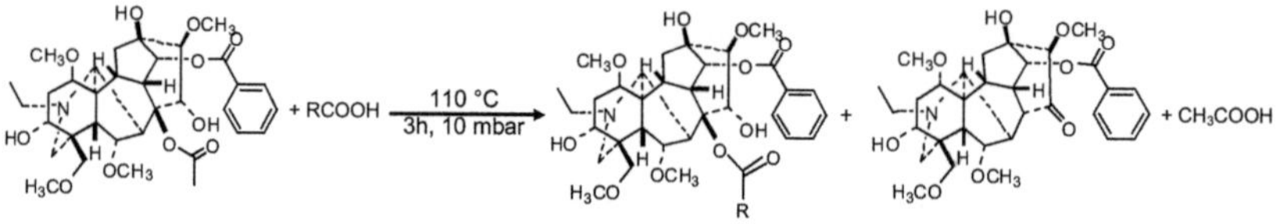

Figure 1. General scheme of the transesterification reaction (R = esterifying fatty acid moiety)

3.2.3 Other alkaloidal compounds

Sporadic occurrences of miscellaneous type alkaloidal compounds have been reported from Delphinieae species. From the aerial parts of *D. fangshanense* atypical occurrence of the tetrahydrobenzylisoquinoline alkaloid, *O*-methylroefractine *N*-oxide and the aporphine alkaloid, magnoliflorine were reported.[34] The latter has also been detected in the aerial parts of *D. pentagynum*.[35] A tetrahydroisoquinoline alkaloid, oleracein E along with two pyrrole alkaloids, aconicaramide and 5-hydroxymethyl-pyrrole-2-carbaldehide, have been isolated from the lateral roots of *A. carmichaelii*. These compounds showed moderate antibacterial activity against several pathogenic bacterial strains, while the latter exerted protective activity against cardiomyocyte damage induced by pentobarbital sodium in primary cultured neonatal rat cardiomyocytes.[36] A new quinazoline alkaloid, 2-(2-methyl-4-oxo-4*H*-quinazoline-3-yl)benzoic acid methyl ester was found in the roots of *A. pseudo-laeve* var. *erectum*.[37] A strange occurrence of a *Spirea* alkaloid, spiratine A has been reported from a Turkish population sample of *Consolida anthoroidea* being the first and until now only report of such a compound in a Ranunculaceae species.[38] Two new amide alkaloids, 3-isopropyl-tetrahydropyrrolo [1,2-a] pyrimidine-2,4 (1*H*,3*H*)-

dione and 1-acetyl-2,3,6-triisopropyl-tetrahydropyrimidine-4(1*H*)-one were isolated from the roots of *A. taipeicum*.[39]

3.2.4 Chemistry of the investigated species

In 2000 MERIÇLI *et al.* reported 6 diterpenoid alkaloids of a Turkish population of *A. anthora* from the EtOH extract of the aerial parts of the plant, C_{20} DAs: isoatisine (**15**), 19-*epi*-isoatisine (**16**), a mixture of 20α-atisine + 20β-atisine (**17**), hetisine (**18**), guan-fu base Y (**19**); C_{19} DA: isotalatizidine (**20**).[40] In a recent study conducted by PIRILDAR *et al.* on another Turkish population of this species guan-fu base A (**21**), condelphine (**22**), nudicaulamine (**23**), and isotalatizidine (**20**) were reported from the

90% EtOH extract of the roots.[41] Italian researchers studied the flavonoids of *A. anthora*. Two new, and two previously known, but from Ranunculaceae species formerly undetected 7,3-*O*-glycosides of quercetin and kaempferol were reported along with the detection of a very modest antioxidant activity of these compounds in three different antioxidant assays.[42]

Preceding our herein detailed work no phytochemical studies have been carried out with *A. moldavicum*.

3.3 Ethnomedicinal uses of *Aconitum* species

European folk medicinal records prove the use of *Aconitum* species as poisons applied even for homicides since the antiquity. On the contrary, Middle and Far Eastern *Aconitum* and *Delphinium* species have a long history of diverse ethnomedicinal uses. In the present thesis only a few characteristic examples are summarised (**Table 1**).

15

Table 1. Ethnomedicinal usage of (non-Chinese) Delphinieae species

Country	Species	Ethnomedicinal use(s)	Ref.
Bhutan	*A. laciniatum*	anthelmintic; chronic infections including leprosy, bone diseases, mumps and gout	[43]
	A. laciniatum, A. violaceum, A. orochryseum	ingredients in traditional medicinal formulas with various indications	[44,45]
India	*A. heterophyllum*	diarrhoea, dysentery, cough, dyspepsia, chronic enteritis; as a febrifuge and bitter tonic in combating debility after malaria	[44]
	D. denudatum	management of nervous disorders and opium addiction	[46]
Japan	*A. carmichaelii*	pain, coldness of distal extremities, recurrent cold episodes, heavy depression, low basal body temperature, vertigo, general fatigue	[47]
	A. carmichaelii, A. japonicum	common cold, polyarthralgia, skin wounds, depression, diarrhoea, heart failure	[48]
Kazakhstan	*A. monticola*	rheumatoid arthritis, neuralgia, deforming osteoarthritis, athetoid spasm	[49]
Kyrgyzstan	*A. karacolicum*	cancer	[50†]
Korea	*A. ciliare*	sinew and joint pain	[51]
	A. chiisanense	neuralgia, arthritis, rheumatism, paralysis, astomia, coldness of extremities, heart disease	[52]
	A. pseudo-laeve var. *erectum*	analgetic and antispasmodic agent; neuralgic and rheumatic conditions	[37]
Mongolia	*Delphinium* spp.	infectious fever, diarrhoea caused by bilious disorder, toothache	[53†]
Nepal	*Delphinium* spp.	rheumatism, cough, fever, toothache; usage recorded even as adulterant to aconites	[54]
	D. scabriflorum	rheumatism and fever, leaf-juice for wound healing	[55]
	A. orochryseum	treatment of fever, diarrhoea, dysentery, cold and cough, tonsillitis, headache, and high altitude sickness	[44]
Turkey	*Consolida* spp., *Delphinium* spp.	analgetics, sedatives, emetics, anthelmintics	[56]
	Consolida spp., *Delphinium* spp.	rheumatic pain, sciatica, body lice	[57,58]
	Delphinium spp.	epileptic seizures, tetanus tremors, rabies, lice	[59]

The fact that besides the two most widely known species (*A. carmichaelii* and *A. kusnezoffii*) literature sources cite that about 76 *Aconitum*, and several *Delphinium* taxa (with the majority being endemic) are currently used in different regions of China[60] (e.g. *A. bulleyanum*[61], *A. coreanum*[62] *A. delavayi*[63], *A. hemsleyanum*[64], *A. nagarum* var. *lasiandrum*[65], *D. densiflorum*[66], *D. trifoliatum*[67] etc.) provides the reason to discuss Chinese usage separately. In China, major and most common applications of drugs from these plants include analgetic, anaesthetic, cardiotonic and antirheumatic indications.[68] They are also frequently recommended for the treatment of kidney disorders and analgesia.[69] In the Tibetan area, extracts of *A. spicatum*,

A. naviculare, and *A. macrorhynchum* are used in analgesic balms, as sedative and febrifuge, and for the treatment of gastricism, hepatitis, and nephritis, respectively.[70,71,72] In the same region, the roots of *A. richardsonianum* var. *pseudosessiliflorum* and the aerial parts of *D. crhysotrichum*[73] have been used for rheumatism and neuralgia.[74] It was reported also from Tibet that for the treatment of infectious fever, pneumonia and inflammation *A. tanguticum*[75,76], and for the treatment of arthralgia, dysmenorrhea, and colic *A. habaense*[77] have long been used as traditional Tibetan medicines.

A very strange culinary tradition has been recorded from different provinces of China, particularly in mountain areas, where people use the roots of certain cultivated aconite species as main ingredient of a tonifying soup predominantly prepared and consumed in the cold season. The soup is boiled for extremely long periods (from 8 hours to two days), and edibility is checked by tasting if numbness of the tongue or the buccal area of the mouth may or may not occur.[4,78]

Concerning Western official applications of aconite drugs *Extractum aconiti* and *Tinctura aconiti* can be mentioned, which were official in the 1st and 2nd editions of the Pharmacopoea Hungarica.[79,80] From the 3rd edition, the Hungarian Pharmacopoeal Commission was among the firsts to omit both the drug and its preparations form the official pharmacopoeia.[81] Interestingly, *Aconiti radix* was included in the French Pharmacopoeia until 1972, in the British Pharmacopoeia until 1973, and it was still official in the Swiss Pharmacopoeia in 1988.[3,82] Currently, there are no *Aconitum* monographs neither in the U.S. Pharmacopoeia (USP) nor in the European Pharmacopoeia (Ph. Eur.). However, the monographs for *Aconiti carmichaelii radix, Aconiti radix praeparata, Aconiti radix lateralis praeparata, Aconiti kusnezoffii radix praeparata,* and *Aconitum napellus ad praeparationes homeopathicas* are under preparation according to the on-line database of Ph. Eur.[83] Official use in Far Eastern countries is provided by the inclusion of both crude and processed drugs of the used species in the respective national pharmacopoeias.

Tuberous roots of *A. carmichaelii* and *A. kusnezoffii* are subjected to various pretreatment methods which result in characteristically different processed drugs. Considering all species used in TCM and the applied processing methods an extensive summary was given by SINGHUBER *et al.*[4] It has to be mentioned that the current (i.e. available in English) 8th edition of the Pharmacopoeia of the People's Republic of China (PhPRC) contains monographs of the crude and processed roots of *A. kusnezoffii* (*Radix aconiti kusnezoffii,* Caowu; *Radix aconiti kusnezoffii praeparata,* Zhicaowu), the crude and processed mother (*Radix aconiti,* Chuanwu; *Radix aconiti praeparata,* Zhichuanwu) and processed daughter roots (*Radix aconiti*

17

lateralis praeparata, Fuzi) of *A. carmichaelii.*[84] This taxon also serves as source of a herbal drug in the 7th edition of the Korean Pharmacopoeia.[85] According to HANUMAN et al.[86†] A. ferox serves the same purpose within the Ayurvedic system.

3.4 Novelties in the pharmacology of diterpene alkaloids

DAs act predominantly on the central nervous and cardiovascular systems; where one of their sites of action is the voltage-gated Na^+ channel of the cell membranes of excitable tissues, on which they act either as activators or as inhibitors.[1†,87†]. Besides this Na^+ ion channel further molecular targets have also been identified, including the NMDA-type glutamate receptor, the A-type GABA receptor, and voltage-gated K^+ channels, just to mention the most important examples.[18,88] Other components acts as antagonists on the nicotinic acetylcholine receptors.[89†] Furthermore, it is well known that some diterpenoid alkaloids have thyrosinase inhibitor, antifeedant and insecticide activities.[90,91] Alkaloids of the genera *Aconitum* and *Delphinium* are of great interest for developing new medicinal preparations because these compounds are known to possess remarkably diverse pharmacological bioactivities on which an extensive list covering the period between 1981 and 2006 has been compiled by SINGHUBER et al.[4]

3.4.1 Cardiac and vascular effects

Although most of the cardiovascular and neurologic effects and poisoning with *Aconitum* products are explained by the activation of Na^+ channels, insufficient evidence is available as concerns other possible mechanisms of action, such as the involvement of the K^+ channels. A previous study reported only on the hERG channel inhibitory effect of aconitine.[92]

Since DAs are known to have different cardiovascular activities, DESAI et al.[93] conducted a study with semisynthetically produced falconerine-8-*O*-stearate and falconerine-8-*O*-linolenate, besides further 12 semisynthetic derivatives. The authors investigated the hypotensive and bradycardic actions of these compounds at two doses (200 μg/kg, 400 μg/kg. iv.) using *N*-deacetyllappaconitine (NDAL) as reference compound by measuring the differences in heart rate and in blood pressure in male Sprague-Dawley rats. It was observed that both compounds had relatively strong activities by reaching, and especially at the dose of 400 μg/kg, surpassing the activities of NDAL. The authors concluded that the examined compounds produced at least 10% fall in heart rate or a 20% decline in blood pressure at both doses. Interestingly, pyrodelphinine was also tested within the same study, but it expressed NDAL-equivalent activity only at the dose of 200 μg/kg, and its activity fell far behind from the activity of NDAL at the dose of 400 μg/kg.

Guan-fu base A, isolated from *A. coreanum* have been developed into a new anti-arrhythmic drug, especially to treat ventricular premature beats and paroxysmal supraventricular tachycardia.[62] This compound, beside lappaconitine hydrobromide[94†], is the second diterpene alkaloid to be applied in the modern medicine as anti-arrhythmic agent.

3.4.2 Effects on the nervous system

Neurobiologically the diterpenoid alkaloids can be divided into three groups. The members of the first group (eg. aconitine, mesaconitine, 3-acetylaconitine) are diesters and are strongly toxic. These compounds activate the voltage sensitive Na^+ channel by linking to the α subunit of the transmembrane glycoprotein. This mechanism is the basis of their antinociceptive effect. The compounds of the second group (eg. lappaconitine, 6-benzoyl-heteratizine, 1-benzoyl-napelline) are monoesters. They are less toxic and are the blockers of the Na^+ channel. Due to this effect they have explicit antinociceptive, antiarrhythmic and antiepileptic effects. The compounds of the third group (eg. heteratisine, napelline, lappaconitine), are not esters. They have no neuronal activity, and experimentally only their antiarrythmic effect was verified.[87†]

Concerning the new results of antinociceptive activities of DAs and aconite drugs it is interesting that this effect was found to be the same in both cases of crude and processed *Radix aconiti lateralis*, while in case of the latter the toxicity as function of aconitine content was found to be reduced.[95] When applying a systems biology approach to detect *Aconitum* alkaloids induced toxicity by mapping the toxic substances into a biological pathway context WANG *et al.* found that aconitine has a direct link with 4 types of ion channels: $Na_{(v)}$ I alpha, SCN 3A ($Na_v1.3$), SCN 2A ($Na_v1.2$), and the tetrodotoxin resistant $Na(I)$ channel, and that the activity towards serotonin, histamine and dopamine receptors is a common bioactivity of *Aconitum* alkaloids.[96] As concerns of the antinociceptive action of *Aconitum* alkaloids it has been revealed that two affinity group of alkaloids can be distinguished: a high and a low affinity group both binding to the Na^+ channel epitope site 2 (without proper specification on the channel subtype). Affinity differences indicate different pharmacological characters in mice (high affinity compounds: induce an increase in synaptosomal $[Na^+]i$ and $[Ca^{2+}]i$ (ED_{50} 3 μM), exert antinociception (ED_{50} 25 μg/kg), provoke tachyarrhythmias, and are highly toxic (LD_{50} 70 μg/kg); low affinity compounds: reduce $[Ca^{2+}]i$, induce bradycardia, exert less antinociception (ED_{50} 50 μg/kg), and are less toxic (LD_{50} 30 mg/kg)), but due to their narrow LD_{50}/ED_{50} values, even the compounds of the low affinity group are deemed unsuitable as analgesics.[4,87†,183†] Guiwuline, a new franchetine-type C_{19} DA isolated from the root bark of *A. carmicahelii* exhibited potential analgesic activity (hot plate test, ED_{50} 15 mg/kg) and little acute toxicity (LD_{50} 500 mg/kg) in a mice, for which the compound can be considered as a lead molecule for the development of novel analgesic agents.[97] Another intensively researched field of CNS-related antinociception effects is the inhibitory effect of processed Aconiti tuber on the development of antinociceptive tolerance to morphine, and its underlying mechanism of the activation of kappa-opioid receptors.[98-102] It has been reported that bulleyaconitine A, a DDA isolated from A. *bulleyanum* is approved for the treatment of chronic pain and rheumatoid arthritis in China. When investigating its underlying effects it was detected that this compound reduces neuronal Na^+ currents in a strong and use-dependent manner that can result in a long-acting local anaesthetic effect, which considering its particular characteristics merits further detailed investigation.[103]

KAWATA *et al.*[104] investigated the conversion of aconitine to lipoaconitines by the human intestinal bacterial flora. The converting activities of *Bacteroides fragilis*, *Klebsiella pneumoniae* and *Clostridium butyricum* were studied by co-incubation with aconitine (**5**), and LAs being esterified with anteiso-pentadeceoic, pentadecenoic, palmitic, palmitoleic, stearic and oleic acids. The esterifying fatty acids were characteristic to the strains used, and the respective LAs were found even after co-incubation of aconitine with sterile bacterial cells or a precipitate of disrupted bacterial cells in a phosphate buffer. As part of this study, antinociceptive activities of 14-benzoylaconine (14-BzA) -8-*O*-palmitate and 14-BzA-8-*O*-oleate were also tested in mice using aconitine as reference compound. Aconitine exerted a significant nociceptive threshold increasing effect with nociceptive hypersensitivity at the dose of 1 mg/kg, but 14-BzA-8-*O*-oleate exhibited this at 3 mg/kg, and it was found to be toxic. 14-BzA-8-*O*-palmitate was active only at the dose of 30 mg/kg, therefore the authors concluded that LAs are not playing an important role in the antinociceptive action of aconite roots.

Further important pharmacological effect that can be related to this group is manifested on the nAChR. Methyllycaconitine, nudicauline and elatine have a selective antagonist effect on the neuronal nAChR at the α-bungarotoxine binding site even in nanomolar concentrations. Experiments made on numerous norditerpenoid alkaloids in order to identify the structure-activity relationship revealed, that the 2-(methyl-succinimido)-benzoyl group is essential for this effect.[89†,105,106†] The outstanding activity of methyllycaconitine on nAChR can bring new therapeutic possibilities in the therapy of Alzheimer's disease.

Different pharmacological activities were reported about C_{20} diterpenoid alkaloids. Peripheral vasodilator effect of kobusine was described in 1997 by

WADA et al.[107] Panicutine, found in *Delphinium denudatum*, has antifungal effect.[108†] Hetisine and related diterpenoid alkaloids were described as repellents and insecticides.[109†,110†,111†]

3.4.3 Antimicrobial activity

DAs isolated from Turkish *Consolida* species, namely lycoctonine, 18-*O*-methyllycoctonine, delcosine, 14-acetyldelcosine and 14-acetylbrowniine presented notable antibacterial effect towards *Klebsiella pneumoniae* and *Acinetobacter baumannii* measured by microdilution method (reference drugs: ampicilline, ofloxacine); quite considerable antifungal activity against *Candida albicans* (reference drug: ketoconazole), and selective and strong inhibitory effect against the RNA virus, *Parainfluenza* (PI-3; reference drugs: acyclovir, oseltamivir) with minimum and maximum cytopathogenic inhibitory concentrations ranging between 1 and 32 μg/ml.[112] Although methanolic extracts of Iranian *Consolida* species previously demonstrated significant activities against *Bacillus subtilis* and *Morganella morganii* (*C. orientalis*), and against *Escherichia coli* and *Candida albicans* (*C. rugulosa*), no compounds exerting these effects were exactly identified or reported.[112] The anti-trypanosomal atisinium chloride isolated from the aerial parts of *A. orochryseum* was found to have moderate antiplasmodial activity against the TM4 (wild) and K1 (chloroquin and antifolate resistant) *Plasmodium falciparum* strains, and because of some *in vivo* toxicity of this alkaloid seen in mice its structure may only be considered as a lead molecule for further research.[44]

3.4.4 Cytotoxic activity

CHODOEVA et al. reported the activity guided isolation 8-*O*-azeoyl-14-benzoylaconine having a truly unique zwitterionic structure and remarkable antiproliferative properties against three human cell lines.[50†] Later, this research group presented a series of mono- and diacylated compounds, differing in the lenght of the alkyl linker, among which compounds containing two aconine moieties interlinked through two ester bonds provided by a pimelate, suberate or azelate at C-11 have been proven to be effective in MCF-7 (breast) and HCT-15 (colon) tumour xenograft bearing *in vivo* mice models with effective doses largely below the maximum tolerated dose.[113]

WADA and HAZAWA investigated natural C_{19} and C_{20} DAs and their analogues against A172 human malignant glioma cells. They found that while the atisine-type alkaloid pseudokobusine did not have antiproliferative properties, its 11-cinnamoate, 11-anisoate and 11-*p*-nitrobenzoate derivatives exerted the highest cytotoxicity activity. They concluded important structure-activity relationships for the cytotoxicity activity.[114][115]

Further alkaloidal compounds that have been reported to show significant *in vitro* cytotoxic activity against various tumor cell lines delelatine (P388, leukaemia)[74]; delphatisine (A549, lung cancer)[73]; aconitine, hypaconitine, mesaconitine and oxonitine (HePG2, liver cancer)[116]; honatisine (MCF-7, breast cancer)[117]; and benzoyldeoxyaconine (HL-60, promyelocytic leukaemia).[118]

3.4.5 Anti-inflammatory and antioxidant activities

CHANG *et al.*[119] reported that the extract of *Radix aconiti* stimulates the secretion of IL-1β, TNF-α and GM-CSF produced by human peripheral mononuclear cells. Although in this study only the extract was studied and no isolated components responsible for the effects were investigated, these results provided the first evidence for pharmacological effects explaining the traditional utilization of aconite drugs in inflammatory conditions. The ethanol extract of the roots of *A. vilmorinianum* used in China as a local substitute for *Aconiti radix* and *Aconiti kusnezoffii radix* was found to induce significant improvement of joint allodynia, swelling, hyperaemia and vascular permeability in an arthritis knee model in the rat.[120] Cochlearenine and lycoctonine, isolated from the roots of *D. linearilobum* were detected to exert significant DPPH radical scavenging and metal chelation, respectively.[121]

3.5 Recent advances in the toxicology of diterpene alkaloids

Due to the fact that aconitine, hypaconitine, and mesaconitine, the major alkaloidal constituents of the most widely applied *Aconitum* species are the most active and the most toxic DDAs (primarily cardio- and neurotoxicity), not just their pharmacology and mode of action, but their toxicological characteristics have also been exhaustively explored.[122†]

The major difference observed in the Asian application of tubers and roots of aconites is that these drugs are used only after cautious processing (usually boiling) to reduce their toxicity. The role of processing in the detoxification of aconite drugs has been and is still being in the focus of attention, which is demonstrated by the mere fact that by 2010 more than 70 methods of investigation have been published.[134] In brief: processing reduces toxicity because of hydrolysis of the ester groups of aconitine-type alkaloids, which results in the formation of less toxic derivatives.[132] First, the acetyl group is hydrolyzed and in the second step the benzoyl group is hydrolyzed (**Figure 2**).

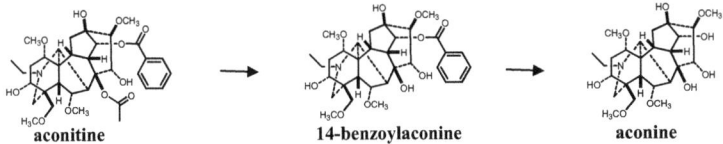

Figure 2. Decomposition of aconitine during processing of *Aconitum* drugs

Because the benzoyl ester is more stable, in the properly processed plant material the concentration of 14-benzoylaconine, 14-benzoylmesaconine, and 14-benzoylhypaconine is remarkable. *Aconitum* roots also contain LAs, which are substantially less toxic than aconitine-type compounds as evidenced by the difference in the lethal intravenous single doses (>10 mg/kg for LAs).[31] A rather alarming possibility of poisoning is that DAs in topically applied aconite preparations are able to permeate through human skin, meaning that in case of such an administration method the risk of developing systemic toxicity is significant, especially with preparations made of unprocessed drugs (e.g. tinctures).[123] From a forensic toxicological/medicolegal point of view it is important to know that according to the regulations stipulated by the Chinese State Food and Drug Administration selling unprocessed *Aconitum* drugs on the Chinese market is forbidden[4], and only processed drugs are authorised for marketing and human use.

With the increasing worldwide spread of traditional oriental healing systems the toxicological issues (including forensic and/or medicolegal cases) raised by incidental or intentional (e.g. suicidal[124]/homicidal[125]) usage of (crude/processed) aconite drugs have gained particular attention and publicity. In Asia, aconite poisoning is much more common because of the continuing and widespread use of aconite roots (unprocessed or processed with reduced DA content)[126†,127-131]; however, with easy access to Asian herbal medicines in Western societies, aconite poisoning can occur anywhere in the world.[132] Although SINGHUBER *et al.* cites several reports on the (Chinese) clinical cases of aconite toxicosis, these cannot be regarded as exhaustive reviewal, but as simple illustrations used to underline the severity of this issue.[4] The number of victims of aconite poisonings reported in the literature varies from the two-digit range to even the four-digit range[4,133-136,*], figuratively culminating around five thousand poisoning cases reported in the past years from China, Hong Kong, Japan, Germany and other countries[143]; and unfortunately, the number of forensic case reports is constantly increasing.[48,137-142] The main causes of death are refractory ventricular arrhythmias and asystole and the overall in-hospital mortality is 5.5%.

---

* It should be noted due to the language barrier imposed by the difficulties in accessing and interpreting the Chinese language literature, these literature sources frequently remain unverifiable for non-Chinese speaking researchers.

23

Amongst the causes of these poisonings not just the insufficiency of (official) control, but the diversity of the alkaloid content of the starting material, improper processing, and last but not least the lack of appropriate quality control can be identified.[136]

Quality control criteria of pharmacopoeias are not always sufficient to warrant safety (as it can be seen in cases of pharmacopeial monographs on *Radix aconiti* and *Radix aconiti kusnezoffii*, in which the PhPRC stipulates no limit for DA contents.) The phenomenon that the usage of these drugs remains a constant source of poisoning urges for better and better methods for controlling the drug processing procedure, and eventually for the assurance of patient safety.[134] This immense need for quality control concerning aconite drugs of human use is highlighted e.g. by a recent investigation on *Radix aconiti lateralis preparata* in which the sum of aconitine, mesaconitine and hypaconitine measured in 8 drug sample types was found to be 3.91-34.8% of that to be found in the corresponding crude drugs.[143] Nonetheless, this latest demonstration of significant variation in the toxic alkaloid contents is in line with practically all other previous studies conducted with the same purpose of DA content comparison in commercial TCM aconite samples.[4] However, it is interesting to note that there are several papers dealing with the detection of the toxic DA alkaloids aconitine, hypaconitine and mesaconitine in human body fluids (urine and blood/plasma samples) by different highly sensitive methods.[132,144-153]

Frequent application of aconite preparation increases the danger of toxicosis. From a comparative investigation using pure aconitine and extract of *Radix aconiti lateralis praeparata* (Fuzi) it was observed that repeated administration only changed the pharmacokinetics of the Fuzi extract with AUC and t_{max} values indicating that multiple administration might increase the bioavailability of aconitine, which in pure form has proven to have low bioavailability, and rapid elimination due to low protein bounding.[154]

Pharmacologic and toxicologic profiles of aconites are influenced by the metabolites of the major alkaloids as well. *In vitro* incubation of the extract of *Radix aconiti* with intestinal bacteria of the rat resulted in the C-8 propionyl, butyryl, an valeryl substituted metabolites of the corresponding DDAs (aconitine, hypaconitine and mesaconitine), which metabolites may be biologically still active. In the liver, aconitine can be transformed into at least six CYP-mediated metabolites in human liver microsomes, and CYP 3A4/5 and 2D6 were found to be the most important CYP isoforms responsible for the de-methylation, *N*-deethylation, dehydrogenation and hydroxylation of this DDA[155]; while besides these enzyme isoforms CYP 2C8/9 and 2D6 also played a minor role in the transformation of mesaconitine into at least nine metabolites following the same processes.[156,157] Results showing reduction of toxicity in an interesting *in vivo* experiment of coadministration of aconitine and paeoniflorin in the rat seems to be supporting the tradition of herbal drug combinations, although, the exact mechanism of this synergistic effect has not been clarified yet.[158] Since data on LD_{50} values (mg/kg body weigh) of the main DAs measured via different administration methods applied in different animal species or observed in human fatalities, and on the acute LD_{50} (mg/kg) values of extracts or compounds of different *Aconitum* species measured in mice are provided comprehensively by SINGHUBER *et al.*[4], and taken into account that CSUPOR summarised the mechanisms of toxicity of DAs and the symptoms of human toxicosis in detail[3], thus rediscussion of these topics is hereby disregarded.

4. MATERIALS AND METHODS

4.1 Plant material

A. anthora was collected at Füzéri Várhegy and Tar-kő in the North Hungarian Mountains in the flowering period in September 2002, and *A. moldavicum* was collected near to Eger in October 2006. The plant material was identified by Attila Molnár V. (Department of Botany, University of Debrecen, Hungary). The plant material of *A. anthora* was dried and stored at room temperature until preparation, while the roots of *A. moldavicum* were freshly ground and extracted. Voucher specimens (numbers: 768 and 807, respectively) are deposited in the Herbarium of Department of Pharmacognosy, University of Szeged.

4.2 Purification and isolation of compounds

For CC, Al_2O_3 (*Al_2O_3 Neutral, Brockman II, Reanal*), for VLC, Al_2O_3 G (*Al_2O_3 60 G neutral, type E, Merck 1090*) was used. Preparative TLC was performed on Al_2O_3 plates (*20 x 20 cm Al_2O_3 60 F_{254}, 1.5 mm, Merck 5715*), and GFC on Sephadex® LH-20 (25-100 µm, Pharmacia Fine Chemicals), or on lipophilic Sephadex® LH20100-100G (Sigma-Aldrich). Purification steps applying CPC were carried out on a Chromatotron® (*Model 8924, Harrison Research*) instrument by using Al_2O_3 G (*Al_2O_3 60 G neutral type E; Merck 1090*) layers. Chromatographic fractions were monitored by TLC on Al_2O_3 plates with the use of cyclohexane–CHCl₃–MeOH (50:50:3), and visualized by spraying with *cc.* H_2SO_4, followed by heating or with Dragendorff's reagent.

4.2.1 *VLC* was carried out on Al$_2$O$_3$; **VLC I**: 108 g; **VLC II**: 20.8 g

VLC I: cyclohexane–CHCl$_3$–MeOH [50:50:0, 50:50:2, 50:50:5, 50:50:10, 50:50:20 and 50:50:50 (350 mL each)], volume of collected fractions: 35 mL.

VLC II: *n*Hex–EtOAc–MeOH [70:30:0, 70:30:0.5, 70:30:1, 70:30:1.5, 70:30:2, 70:30:3, and 70:30:5, (100 mL each)], volume of collected fractions: 10 mL.

4.2.2 *CC* was performed on Al$_2$O$_3$; **CC I**: 1428.5 g

CC I: cyclohexane–EtOAc–MeOH [70:30:0, 70:30:1, 70:30:1.5, 70:30:2, 70:30:3, 70:30:5, 70:30:7, 70:40:10, 60:40:20 and 60:40:50 (3000, 6900, 2700, 1800, 3000, 2400, 2400, 1800, 2700 and 5100 mL)], volume of collected fractions: 300 mL.

4.2.3 *PLC* was carried out on Al$_2$O$_3$. Separation was monitored in UV light at 254 nm or by spraying the border of the plates with Dragendorff's reagent. Compounds were eluted with CHCl$_3$–MeOH 1:1. Mobile phases:

PLC I: cyclohexane–CHCl$_3$–MeOH 50:30:1

PLC II-V: To–acetone–EtOH–*cc.* NH$_3$ 70:40:10:3

4.2.4 *GFC* was performed on Sephadex® LH-20; **GFC I-IX**: 5.0 g each. Number of collected fractions: 20. Volume of collected fractions: 1 and 20: 20 mL each, further fractions 1 mL each. Mobile phases:

GFC I-III, V-VIII: acetone

GFC IV, IX: CHCl$_3$–MeOH 1:1

4.2.5 *CPC* was carried out with a Chromatotron® on manually prepared Al$_2$O$_3$ plates, thickness 1-4 mm, flow rate 4-10 mL/min. Mobile phases:

CPC I: CH$_2$Cl$_2$–MeOH [98:2, 97:3, 95:5, 93:7, 90:10, and 80:20 (25, 45, 45, 45, 50, 25 mL, respectively)], volume of collected fractions: 5 mL.

CPC II: *n*Hex–EtOAc–MeOH [70:30:3, 70:30:3.5, 70:30:4, 70:30:4.5, and 70:30:5, (50 mL each)], volume of collected fractions: 5 mL.

CPC III, VI: *n*Hex–CH$_2$Cl$_2$–MeOH [70:30:1, 70:30:2, 70:30:3, 70:30:4, 70:30:5, and 70:30:7, (50 mL each)], volume of collected fractions: 5 mL.

CPC IV-V: *n*Hex–CH$_2$Cl$_2$–MeOH [70:30:3, 70:30:4, 70:30:5, and 70:30:7 (50 mL each)], volume of collected fractions: 5 mL.

CPC VII: To–acetone–EtOH–*cc.* NH$_3$ [70:40:10:1, and 70:40:10:2 (40 mL each)], volume of collected fractions: 2 mL.

4.3 Semisynthesis of lipo-alkaloids

The semisynthesis of aconitine-derived LAs was carried out according to the modified method of BAI *et al.*[159] as elaborated by CSUPOR *et al.*[31] In brief the reaction mixtures were heated in an oil bath (110 °C, except of esterification with eicosapentaenoic and docosahexaenoic acids, for which the temperature was lowered to 90 °C, due to the presumed higher heat-sensitivity of these compounds) for 3 h under vacuum (10 mbar). In the reactions 20 mg aconitine (**5**) was esterified by 40 mg lauric, myristic, stearic, palmitoleic, oleic, α- and γ-linolenic, eicosanoic, 11Z-eicosenoic, 11Z,14Z-eicosadienoic, 8Z,11Z,14Z-eicosatrienoic, eicosapentaenoic acids, and 25 mg docosahexaenoic acid, respectively. All chemical substances used for the reactions were purchased as highly purified test reagents. Supplier for all compounds was Sigma-Aldrich Ltd., Hungary. For purity (catalogue number) of compounds see refs. [19] and [160]. To obtain the pure LAs and the by-product, pyroaconitine (**24**), the reaction mixtures were successively purified by gelfiltration, preparative TLC and CPC as described in refs. [160] and [19]. The identity and purity of the compounds were investigated by ^{1}H- and ^{13}C-NMR spectroscopy.

4.4 Characterisation and structure elucidation

Optical rotations were determined in CHCl$_3$ at room temperature with a Perkin-Elmer 341 polarimeter. NMR spectra were recorded in CDCl$_3$ on a Bruker Avance DRX 500 spectrometer, at 500 MHz (^{1}H) and 125 MHz (^{13}C), with TMS as internal standard. Two-dimensional data were acquired and processed with standard Bruker software. In the ^{1}H-^{1}H COSY, HSQC and HMBC experiments, gradient-enhanced versions were used. HRMS measurements were performed on a Finnigan MAT 95 S and a VG ZAB SEQ hybrid mass spectrometer equipped with a Cs SIMS ion source. Melting points are uncorrected.

4.5 Pharmacological tests carried out with isolated and semisynthesised alkaloids

4.5.1 Assays for COX-1, COX-2 and LTB$_4$ formation inhibitory activity

For detailed protocols of assays for COX-1, COX-2 and LTB$_4$ formation inhibitory activities see ref. [160] In the inhibition assays for COX-1 and COX-2 indomethacine (COX-1, IC$_{50}$ 0.9 μM) and NSB-398 (COX-2, IC$_{50}$ 2.6 μM) were used as positive controls. In the leucotriene B$_4$ formation inhibition assay zileuton (IC$_{50}$ 5.0 μM, Sigma Aldrich) was used as positive control.[160]

4.5.2 Alkaloids tested in the bioassays of hERG and Na$_v$1.2 channels

Semisynthetic preparations of 14-BzA-8-*O*-palmitate (**48**) and pyroaconitine (**24**), alike the isolation of acotoxicine (**1**), aconosine (**49**), dolaconine (**50**), delectinine (**51**), neolinine (**52**), neoline (**53**), acotoxinine (**54**), songoramine (**55**) and songorine (**56**) from *A. toxicum;* acovulparine (**57**) and septentriodine (**58**) from *A. vulparia;* delcosine (**40**), gigactonine (**44**), takaosamine (**59**) and 14-desacetyl-18-

demethylpubescenine (**60**) from *Consolida orientalis* was described in ref. [161] (for structures see **Annex II**). Aconitine (**5**) was purchased from Sigma-Aldrich. The purities of the compounds were investigated by means of ^{1}H-NMR spectroscopy[162], and all found to be >95%. In the hERG bioassay haloperidol was used as a positive control (Sigma).

Whole-cell patch clamp analysis of the effects on hERG channels

CHO cells stably expressing the transcript of hERG were investigated by the automated whole-cell patch clamp technique, using the QPatch-16 system (Sophion). For detailed protocol of hERG bioassay see ref. [161]

Whole-cell patch clamp analysis of current amplitude modulation

CHO cells stably expressing human $Na_v1.2$ sodium channels were investigated by the whole-cell patch clamp using the QPatch-16 automated patch clamp system. Cells were held at -65 mV and activated by a train consisting of 20 steps to 0 mV for 8 ms at 10 Hz. Train protocols were applied at 10-s-intervals. The current traces were recorded, and the amplitude of peak current evoked by the last pulse within a train was measured as test parameter. The inhibition was calculated from the peak currents in the presence and absence of the test compound. The control solution contained the same concentration of the vehicle (DMSO or HCl) as the solutions of the test compound. Solutions containing the test compounds were applied to the cells for 6-8 min.

4.6 Determination of toxic alkaloid contents of processed *Radix aconiti* samples

Processed *Aconitum* samples were obtained from different suppliers (**Table 5**). Unprocessed *A. carmichaelii* roots were obtained from a pharmacy in China. Aconitine, mesaconitine, and hypaconitine were purchased from PhytoLab GmbH.[82]

For HPLC, an improved version of a recently reported method has been applied.[82] Peaks of mesaconitine, aconitine, and hypaconitine were identified by comparison of the HPLC-DAD chromatograms of the extracts of aconite roots with those of the reference solutions. Alkaloid content was calculated by comparison of the sum of the areas under curves (AUC) of mesaconitine, aconitine, and hypaconitine on the basis of the calibration curve established for aconitine. Calibration was established for aconitine based on five concentrations (with a range of 0.05–1.625 µg).[82]

Alkaloid titration was carried out according the method of the *German Homeopathic Pharmacopoeia* as published in ref. [82]

5. RESULTS

5.1 Isolation of alkaloids

5.1.1 Isolation of alkaloids from A. anthora

The dry herbal sample was 220 g. The ground sample was percolated with MeOH (12 L) (**Figure 3**). After evaporation *in vacuo*, the concentrated extract (370 mL) was diluted with water (130 mL), and acidified with 4% H_2SO_4 (250 mL). After the removal of neutral materials with $CHCl_3$ (5×500 mL), the acidic solution was adjusted to pH 9.0 with 5% NaOH, and extracted with $CHCl_3$ (5×500 mL) to yield the crude alkaloid fraction (3.09 g). This crude fraction was first separated by VLC (**VLC I**). Fractions eluted with cyclohexane–$CHCl_3$–MeOH (50:50:2) having similar composition (VLC I/6-10) were subjected to preparative LC (**PLC I**) yielding **ANT-1** (12 mg, amorphous solid). Fractions obtained with cyclohexane–$CHCl_3$–MeOH (50:50:2) (VLC I/17-19) were subjected to repeated gel chromatography (**GFC I** and **GFC II**). The obtained compound was combined with the main alkaloid constituent obtained with consequent GFC (**GFC III**) of VLC I fractions 20-23, then further purified with and PLC (**PLC II**). This process afforded **ANT-2**, (8.2 mg, amorphous solid). The fractions obtained from VLC I with the solvent system cyclohexane–$CHCl_3$–MeOH 50:50:5 (VLC I/24-27) were subjected to GFC (**GFC IV**). Upon standing, the alkaloid fraction of this separation furnished a crystalline material, which was recrystallised from MeOH, yielding **ANT-3** (23.0 mg, m.p. 283-5 °C) in pure form.

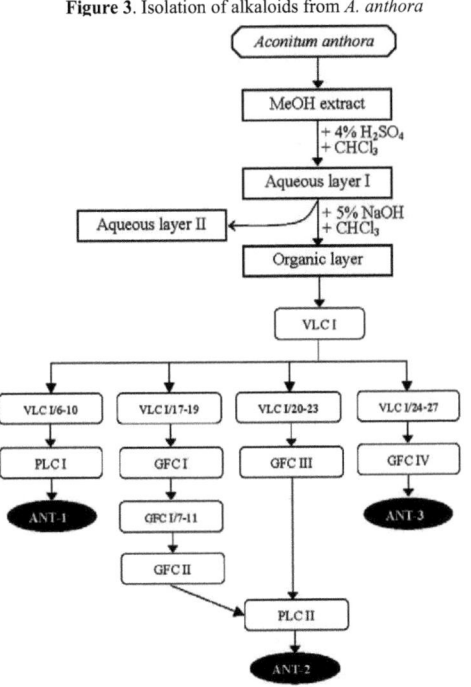

Figure 3. Isolation of alkaloids from *A. anthora*

5.1.2 Isolation of alkaloids from A. moldavicum

The fresh sample mass was 1086 g. The ground sample was percolated with 15.2 L MeOH and, after evaporation *in vacuo*, the concentrated, dry residue was 40.6 g. This residue was separated by CC (**CC I**) on Al_2O_3 (1428.5 g), using a gradient system of cyclohexane–EtOAc–MeOH with increasing polarity (**Figure 4**). Crystalline material precipitated from fraction 10 (CC I/10) eluted with cyclohexane–EtOAc–MeOH (70:30:1). Recrystallisation was carried out using cyclohexane–EtOAc 7:3, yielding **AMO-2** (122 mg, m.p. 136-9 °C) in pure form.

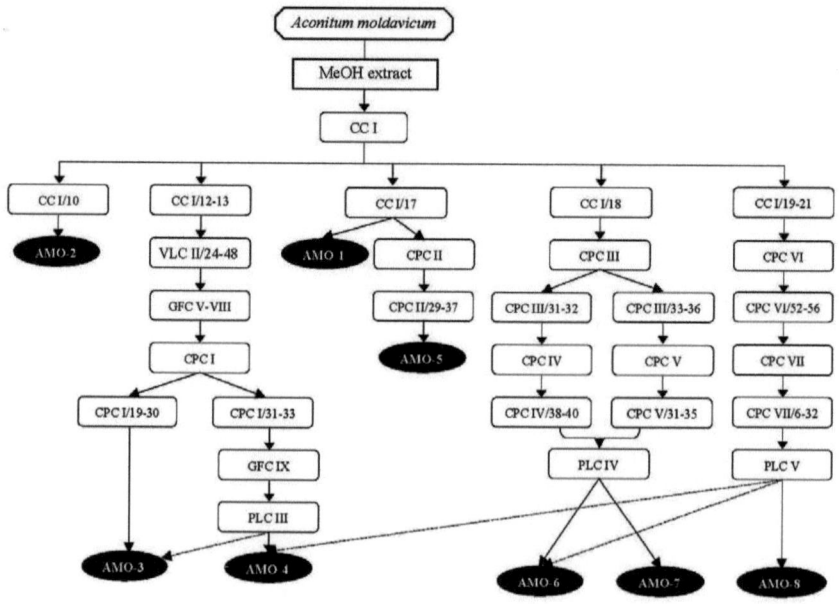

Figure 4. Isolation of alkaloids from *A. moldavicum*

Fractions eluted with cyclohexane–EtOAc–MeOH (70:30:1) having similar alkaloid composition (CC I/12-13) were combined and subjected to VLC (**VLC II**) using a gradient system of *n*Hex–EtOAc–MeOH with increasing polarity. Fractions eluted with *n*Hex–EtOAc–MeOH (70:30:1, 70:30:1.5, and 70:30:2) (VLC II/24-48) were further purified with repeated gel chromatography (**GFC V-VIII**). The resulting fractions having similar composition were combined and further purified by CPC on Al_2O_3 stationary phase with a gradient system of CH_2Cl_2–MeOH with increasing polarity (**CPC I**). Fractions eluted from CPC I with CH_2Cl_2–MeOH 95:5 and 93:7 (CPC I/19-30) furnished crystalline **AMO-3** (214 mg, m.p. 89-94 °C) in pure form.

Fractions eluted from CPC I with CH_2Cl_2–MeOH 93:7 (CPC I/31-33) were further purified with gel chromatography (**GFC IX**) followed by PLC (**PLC III**) to obtain **AMO-4** (2.9 mg, amorphous solid) along with further amount of lycoctonine (**AMO-3**, 22.8 mg). Crystallisation was also observed in case of fraction 17 eluted from CC I with cyclohexane–EtOAc–MeOH (70:30:3) (CC I/17). The obtained material was recrystallised from cyclohexane–EtOAc 7:3 to yield **AMO-1** (753 mg, m.p. 201-5 °C) in pure form. The mother liquor of this fraction was subjected to CPC on Al_2O_3 stationary phase with a gradient system of nHex–EtOAc–MeOH with increasing polarity (**CPC II**). Fractions eluted from CPC II with nHex–EtOAc–MeOH 70:30:4 and 70:30:4.5 (CPC II/29-37) furnished **AMO-5** (27 mg, amorphous solid). Fraction eluted from CC I with cyclohexane–EtOAc–MeOH (70:30:3) (CC I/18) was further purified at first instance by CPC on Al_2O_3 stationary phase with a gradient system of nHex–CH_2Cl_2–MeOH with increasing polarity (**CPC III**). Subfractions eluted from CPC III with nHex–CH_2Cl_2–MeOH (70:30:4) (CPC III/31-32, CPC III/33-36) were parallely subjected to CPC on Al_2O_3 stationary phase with a gradient system of nHex–CH_2Cl_2–MeOH with increasing polarity (**CPC IV** and **CPC V**, respectively). Respective subfractions eluted with nHex–CH_2Cl_2–MeOH (70:30:7) (CPC IV/38-40, CPC V/31-35) were combined and then finally purified using PLC (**PLC IV**) to obtain **AMO-6**, (2 mg, amorphous solid) along with **AMO-7** (3.6 mg, amorphous solid). Combined fractions eluted from CC I with cyclohexane–EtOAc–MeOH (70:30:5 and 70:30:7) (CC I/19-21) were purified by CPC on Al_2O_3 stationary phase with a gradient system of nHex–CH_2Cl_2–MeOH with increasing polarity (**CPC VI**). Subfractions eluted from CPC VI with nHex–CH_2Cl_2–MeOH (70:30:7) (CPC VI/52-56) were again subjected to CPC on Al_2O_3 stationary phase using now a two-step gradient of To–acetone–EtOH–$cc.$ NH_3 (**CPC VII**). Fractions obtained from this step (CPC VII/6-32) were combined and finally purified by using PLC (**PLC V**) to yield the new C_{19} DA **AMO-8** (2 mg, amorphous solid) along with further amounts of **AMO-4** (25.8 mg) and **AMO-6** (7.4 mg)

5.2 Structure elucidation of the isolated compounds

5.2.1 Alkaloids from A. anthora

ANT-2

ANT-2 was isolated as amorphous solid; $[\alpha]_D^{28}$: −7 (c 0.1, $CHCl_3$); for ^{1}H- and ^{13}C-NMR data see **Table A in Annex IV**. The ESI-MS investigation demonstrated the presence of a molecular ion peak at m/z 452.2994 $[M+H]^+$ ($C_{25}H_{42}NO_6$), which afforded fragment ions at m/z 420.2741 $[M+H–CH_3OH]^+$, 388.2476 $[M+H–2xCH_3OH]^+$ and 370.2372 $[M+H–2xCH_3OH–H_2O]^+$. The ^{1}H-NMR and

JMOD spectra of **ANT-2** showed signals at δ_H 1.08 t, 2.50 m and 2.41 dq and δ_C 49.4 and 13.6, characteristic of an *N*-ethyl group. Singlet signals at δ_H 3.14, 3.28, 3.31 and 3.37 (each 3H, *s*) and carbon signals at δ_C 48.2, 56.0, 56.4 and 59.5 were indicative of the presence of four methoxy groups. In addition, the JMOD spectrum demonstrated that **ANT-2** has a C_{19} norditerpene core, composed of four quaternary carbons, eight methines and seven methylenes (**Table A in Annex IV**).

The HSQC experiments allowed the assignment of protons and protonated carbons. Interpretation of the proton-proton connectivities in the ^{1}H-^{1}H COSY spectrum led to the identification of three structural fragments: -CH-CH$_2$-CH$_2$- (unit A, C-1−C-3), -CH-CH$_2$-CH-CH- (unit B, C-5−C-6−C-7−C-17) and -CH-CH(OR)-CH(CH$_2$-)-CH(OR)-CH$_2$- (unit C, C-9−C-14−C-13(C12)−C-16−C-15), two isolated methylenes [δ_H 3.01 d, 3.13 d (C-18) and 2.02 d, 2.54 d (C-19)], and one *N*-ethyl group. The long-range correlations form HMBC confirmed, that structural fragments A-C and the two methylene groups (C-18 and C-19), together with the quaternary carbons C-4, C-8, C-10 and C-11, build up a norditerpene skeleton. Important two- and three-bond correlations were observed between the quaternary carbon at δ_C 54.8 (C-11) and the protons at δ_H 3.75 (H-1), 2.41 (H-7), 2.66 (H-12a) and 2.87 (H-17), and between the carbon at δ_C 38.4 (C-4) and the protons at δ_H 1.75 (H-3a), 1.42 (H-3b), 1.84 (H-5), 3.13 (H-18a), 3.01 (H-18b), 2.54 (H-19a) and 2.02 (H-19b), indicating that fragment A and the quaternary carbons C-4 and C-11 form a six-membered ring. The correlations of the quaternary carbon at δ_C 77.0 (C-8) with the protons at δ_H 1.82 (H-6a), 1.47 (H-6b), 2.41 (H-7), 1.99 (H-9), 2.19 (H-15a) and 2.10 (H-15b), and of the carbon at δ_C 81.1 (C-10) with the protons at δ_H 3.75 (H-1), 1.99 (H-9), 2.66 (H-12a), 1.71 (H-12b), 2.50 (H-13) and 2.87 (H-17) proved the linkage of structural parts C and B as in the aconitane skeleton with six oxygen functions. The HMBC spectrum also provided information on the locations of the four methoxy groups. The long-range correlations of C-1, C-8, C-16 and C-18 with the protons at δ_H 3.28, 3.14, 3.37 and 3.31 (each 3H) revealed the positions of methoxy groups at C-1, C-8, C-16 and C-18, respectively. The 8-OMe group displayed an unusual high-field carbon resonance (δ_C 48.2), similarly as observed for other 8-methoxy-substituted NDAs.[163] The two hydroxy groups in **ANT-2**, which were evident from the molecular mass and the carbon resonances at δ_C 73.3 and 81.1, were located of necessity on C-10 and C-14.

The relative stereochemistry of **ANT-2** was elucidated by analysing the NOESY spectrum. As starting point, the stereochemistry of H-5 was considered to be β, as characteristic for NDAs. The *Overhauser* effects between H-5

10-hydroxy-8-O-methyltalatizamine
(**ANT-2**)

and H-1, and between H-5 and H-18a, were indicative of the β position of H-1 and the 18-methylene group. The NOESY correlations between H-19 and H-17, H-19 and H-21, H-17 and H-16, and H-17 and H-12α pointed to α-oriented protons, and corroborated that the *N*-containing bridge, between C-19 and C-17, including the *N*-ethyl group, is below the plane of the six-membered ring. Further important NOEs were detected between H-9 and H-14, H-13 and H-14, H-12β and H-13, H-12β and H-14, H-13 and 16-OMe, and H-15β and 8-OMe, which confirmed the β orientation of all of these protons and groups. Furthermore, some NOESY correlations were suitable for the steric differentiation of methylene protons (H₂-2, H₂-3, H₂-6, H₂-12 and H₂-15). Such correlations were observed between H-1 and H-2β; H-1 and H-3β; H-2α and H-3α; H-9 and H-6β; H-16 and H-12α; H-16 and H-15α; H-17 and H-15α; H-15β and 8-OMe; H-13 and H-12β; and H-14 and H-12β. All of the above evidence confirmed the structure of **ANT-2** as 10-hydroxy-8-*O*-methyltalatizamine.

ANT-1

The ¹H-NMR spectrum of **ANT-1** showed the presence of two methoxy [δ 3.32 and 3.34 ppm (2x3H, 2xs)] and one *N*-ethyl groups [δ 1.12 ppm (3H, t, *J* = 7,0 Hz)]. The JMOD spectrum of this compound proved the presence of two methoxy groups (δ 56.3, 59.4 ppm) and an *N*-ethyl substituent (δ 48.5, 13.0 ppm). In addition, it indicated further 19 carbon atoms, namely a norditerpenoid core. The carbon skeleton is built up by 9 methine and 10 methylene+quaternary carbon atoms, distinguished them on the basis of the ¹³C resonance signals. Such structure corresponds to isotalatizidine (**20**). The spectral data were compared, and it was concluded that NMR data of the isolated compound and of isotalatizidine (**Table C in Annex IV**) showed good correlation, thus, **ANT-1** is identical with isotalatizidine (**20**).

isotalatizidine
(ANT-1)

This compound was reported earlier from a Turkish population of *A. anthora* by MERIÇLI *et al.*[40] We found ¹³C-NMR chemical shifts identical to those described earlier,[164] but our 2D NMR investigations, including ¹H-¹H COSY, HSQC and HMBC experiments, permitted some revised ¹³C assignments and complete chemical shift assignments for all protons of **20** (**Table C in Annex IV**). The HMBC correlations between C-3/H-1, C-2/H₂-3, C-3/H₂-18, C-3/H₂-19, C-10/H-1, C-10/H-5, C-10/H-9, C-10/H₂-12, C-10/H-13, C-10/H-17, C-12/H-9, C-12/H-10, C-12/H-16, C-13/H-9, C-13/H-12, C-13/H-14, C-13/H-15, and C-13/H-16 indicated the ¹³C reassignments of C-2, C-3, C-10, C-12 and C-13, as listed in **Table C in Annex IV**.

ANT-3

ANT-3 was isolated as white crystals (m.p. 283-5 °C). Its JMOD spectrum contained signals of 20 carbon atoms, which allows to conclude a C_{20} diterpenoid structure. Signals at 145.3 and 108.1 ppm were indicative to an exomethylene group in the molecule, which is common in the C_{20} diterpenoid alkaloids isolated from *Aconitum* species. Additional ^{13}C-NMR signals showed the presence of 9 secondary+quaternary, 8 methine and 1 methyl groups. The carbon signal at 212.9 ppm refers to a keto group. In the ^{1}H-NMR spectrum of **ANT-3** 1 methyl signal connectioned to a quaternary carbon atom was identified (δ 1.14 s, 3H). Additionally, hydrogens of the exomethylene group could be detected at δ 4.86 (s) and 4.68 (s) ppm. It was found that the ^{1}H and ^{13}C chemical shifts of **ANT-3** are in good agreement with the data of hetisinone (**39**) published by DE LA FUENTE *et al.* The ^{13}C chemical shifts correlated well with the data published by GONZALEZ *et al.* The above evidence indicated that **ANT-3** is identical with hetisinone (**39**).[165,166]

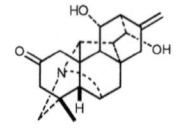

hetisinone
(ANT-3)

5.2.2 Alkaloids from A. moldavicum

AMO-8

AMO-8 was isolated as amorphous solid; $[\alpha]_D^{28}$: +22 (*c* 0.1, CHCl₃); for ^{1}H- and ^{13}C-NMR data see **Table B in Annex IV**. It was shown by HRESIMS to have the molecular formula of $C_{24}H_{39}NO_8$ according to the quasimolecular ion peak at *m/z* 470.2761 [M+H]$^+$ (calcd. for $C_{24}H_{40}O_8$ 470.2754), which afforded fragment ions at *m/z* 452 [M+H−H₂O]$^+$ and 320 [M+H−CH₃OH]$^+$. The ^{1}H and ^{13}C JMOD spectra of **AMO-8** indicated the presence of an *N*-ethyl group (δ_H 1.11 t, 2.97 m and 2.85 dq; δ_C 50.3 and 14.0). Singlet signals at δ_H 3.35, 3.43 and 3.45 (each 3H, *s*) and carbon signals at δ_C 56.3, 57.8 and 58.0 demonstrated the presence of three methoxy groups. Further, the JMOD spectrum suggested that the skeleton consists of 19 carbons, including six methylenes, eight methines and five quaternary carbons (**Table B in Annex IV**). From the HSQC spectrum, the chemical shifts of the protonated carbons were assigned, and the proton-proton connectivities were then studied. The ^{1}H-^{1}H COSY spectrum defined structural fragments with correlated protons: −CHR-CH₂-CH₂− (unit A, C-1−C-3) (δ_H 4.06 brs, 1.72 m, 1.68 m, 1.98 m and 1.47 m), −CH−CHR− (unit B, C-5−C-6) (δ_H 2.15 s, 4.03 s), −CH−CHR−CH−CH₂− (unit C, C-9−C-14−C-13−C-12) (δ_H 2.83 d, 4.11 t, 2.58 dd, 2.34 d and 1.95 dd), −CH₂−CH− (unit D, C-15−C-16) (δ_H 2.63 dd, 1.80 dd and 3.26 dd), two isolated methylenes [δ_H 3.40 d, 3.70 d (C-18) and 2.45 d, 2.49 d (C-19)], and one *N*-ethyl group. The long-range correlations detected in the HMBC spectrum proved, that structural elements

A-D and the two methylene groups (C-18 and C-19), together with the quaternary carbons C-4, C-7, C-8, C-10 and C-11, build up an aconitane diterpene substituted in C-1, 6, 7, 8, 10, 14, 16, 18 positions (**Table B in Annex IV**). The location of the methoxy groups were established via HMBC experiments. The long-range correlations of C-6, C-14 and C-16 with the protons at δ_H 4.03 s, 4.11 t and 3.26 dd demonstrated the presence of methoxy groups at C-6, C-14 and C-16, respectively. The four hydroxy groups in **AMO-8** were located of necessity on C-1, C-7, C-8, C-10 and C-18. The stereochemistry and relative configuration of **AMO-8** were studied by means of a NOESY experiment (**Table B in Annex IV**). As reference point, the β stereochemistry of H-5 was used, which is characteristic for NDAs. The *Overhauser* effects between H-5/H-9, H-5/H-3β, H-3β/H-1, H-9/H-14, H-9/8-OH, H-14/13, H-13/H-12β, H-12β/H-14 and 8-OH/H-6 were indicative of the β position of these protons and 8-OH group. On the other hand, NOE effects observed between H-6/H-19, H-17/H-16, H-17/H-20, H-16/H-12α, H-16/H-15α exhibited the α-orientation of H-6, H-16 and corroborated that the *N*-containing bridge, between C-19 and C-17, including the *N*-ethyl group, is below the plane of ring A. All of the above evidence was used to propose the structure of this compound as new alkaloid, 1-*O*-demethylswatinine (**47**).

1-O-demethylswatinine
(AMO-8)

Further compounds isolated from *A. moldavicum* were identified as delcosine (**40, AMO-1**), ajacine (**41, AMO-2**), lycoctonine (**42, AMO-3**), gigactonine (**44, AMO-5**), and cammaconine (**45, AMO-6**) by comparing their spectral data with those reported in the literature.[167,168,169] Swatinine (**43, AMO-4**) and columbianine (**46, AMO-7**) are known compounds of the *Aconitum* genus (found in *A. laeve, A. ferox* and *A. lamarckii*)[91,170,171†], but our 2D NMR studies provided the complete ^{1}H assignments for these alkaloids for the first time (**Table D in Annex IV**).

delcosine
(AMO-1)

ajacine
(AMO-2)

lycoctonine
(AMO-3)

swatinine
(AMO-4)

gigactonine
(AMO-5)

cammaconine
(AMO-6)

columbianine
(AMO-7)

5.3 Semisynthesis of lipo-alkaloids

As a result of the semisynthesis the following transesterified compounds were obtained: 14-BzA-8-*O*-laurate (**25**), 14-BzA-8-*O*-myristate (**26**), 14-BzA-8-*O*-stearate (**27**), 14-BzA-8-*O*-palmitoleate (**28**), 14-BzA-8-*O*-oleate (**29**), 14-BzA-8-*O*-α-linolenate (**30**), 14-BzA-8-*O*-γ-linolenate (**31**), 14-BzA-8-*O*-eicosanoate (**32**), 14-BzA-8-*O*-eicosa-11*Z*-enoate (**33**) 14-BzA-8-*O*-eicosa-11*Z*,14*Z*-dienoate (**34**), 14-BzA-8-*O*-eicosa-8*Z*,11*Z*,14*Z*-trienoate (**35**) 14-BzA-8-*O*-eicosapentaenoate (**36**), 14-BzA-8-*O*-docosahexaenoate (**37**).

R: esterifying fatty acid

25-37

Comp. No.	Esterifying fatty acid	Corresponding lipo-alkaloid (BzA = benzoylaconine)
25	lauric	14-BzA-8-*O*-laurate
26	myristic	14-BzA-8-*O*-myristate
27	stearic	14-BzA-8-*O*-stearate
28	palmitoleic	14-BzA-8-*O*-palmitoleate
29	oleic	14-BzA-8-*O*-oleate
30	α-linolenic	14-BzA-8-*O*-α-linolenate
31	γ-linolenic	14-BzA-8-*O*-γ-linolenate
32	eicosanoic	14-BzA-8-*O*-eicosanoate
33	11*Z*-eicosenoic	14-BzA-8-*O*-eicosa-11*Z*-enoate
34	11*Z*,14*Z*-eicosadienoic	14-BzA-8-*O*-eicosa-11*Z*,14*Z*-dienoate
35	8*Z*,11*Z*,14*Z*-eicosatrienoic	14-BzA-8-*O*-eicosa-8*Z*,11*Z*,14*Z*-trienoate
36	eicosapentaenoic	14-BzA-8-*O*-eicosapentaenoate
37	docosahexaenoic	14-BzA-8-*O*-docosahexaenoate

5.4 Pharmacological tests with the isolated and semisynthesised alkaloids

5.4.1 Assays for COX-1, COX-2 and LTB_4 formation inhibitory activity

The aim of the semisynthetic preparation of a LA series was to gain information about the pharmacology of these compounds and to conclude some structure-activity relationships concerning the nature of esterifying fatty acids.[160] *In vitro* anti-inflammatory activities of altogether thirteen LAs (**25–37**) were evaluated by COX-1, COX-2 and LTB_4 formation inhibitory assays (**Table 2**).

Table 2: *In vitro* anti-inflammatory activities of semisynthetic LAs

Compound	Inhibition % ± SD*		
	COX-1	COX-2	LTB$_4$ formation
14-BzA-8-*O*-laurate (**25**)	-29.8 ± 5.7	-8.0 ± 5.6	33.5 ± 11.2
14-BzA-8-*O*-myristate (**26**)	-25.3 ± 21.6	-4.0 ±15.1	35.9 ± 9.8
14-BzA-8-*O*-stearate (**27**)	-33.5 ± 39.3	-3.5 ± 10.2	52.5 ± 5.9
14-BzA-8-*O*-palmitoleate (**28**)	-18.8 ± 8.7	17.1 ± 9.0	34.5 ± 6.5
14-BzA-8-*O*-oleate (**29**)	-24.3 ± 17.0	15.3 ± 28.9	45.5 ± 4.6
14-BzA-8-*O*-α-linolenate (**30**)	-5.9 ± 20.7	22.0 ± 7.8	48.0 ± 1.9
14-BzA-8-*O*-γ-linolenate (**31**)	-11.8 ± 22.8	12.7 ± 4.9	34.7 ± 1.7
14-BzA-8-*O*-eicosanoate (**32**)	25.7 ± 6.5	-1.8 ± 22.1	30.9 ± 11.8
14-BzA-8-*O*-eicosa-11*Z*-enoate (**33**)	25.1 ± 2.2	-19.7 ± 9.1	25.7 ± 6.5
14-BzA-8-*O*-eicosa-11*Z*,14*Z*-dienoate (**34**)	26.3 ± 10.8	34.7 ± 17.5	25.1 ± 2.2
14-BzA-8-*O*-eicosa-8*Z*,11*Z*,14*Z*-trienoate (**35**)	29.9 ± 2.3	no data available	26.3 ± 10.8
14-BzA-8-*O*-eicosapentaenoate (**36**)	54.5 ± 24.4	66.1 ± 3.5	46.0 ± 3.3
14-BzA-8-*O*-docosahexaenoate (**37**)	15.1 ± 24.2	40.2 ± 8.8	61.0 ± 7.9

* average of 2 tests in duplicate, the compounds were tested at 50 μM concentration.
IC$_{50}$ values of positive controls: indomethacine (COX-1): 0.9 μM; NSB-398 (COX-2): 2.6 μM; and zileuton (LTB$_4$ formation inhibition) 5.0 μM

In the COX-1 inhibition assay 14-BzA-8-*O*-eicosapentaenoate (**36**) exhibited the highest activity with 54.5 % inhibition at 50 μM. Besides this compound, 14-BzA-8-*O*-eicosanoate (**32**), 14-BzA-8-*O*-eicosa-11*Z*-enoate (**33**) 14-BzA-8-*O*-eicosa-11*Z*,14*Z*-dienoate (**34**), 14-BzA-8-*O*-eicosa-8*Z*,11*Z*,14*Z*-trienoate (**35**), and 14-BzA-8-*O*-docosahexaenoate (**37**) showed moderate inhibitory activities, all other LAs (**25-31**) were inactive against the COX-1 enzyme.

In the COX-2 inhibition assay, also 14-BzA-8-*O*-eicosapentaenoate (**36**) demonstrated remarkable activity (66.1% inhibition at 50 μM), which was followed by the moderately effective 14-BzA-8-*O*-eicosa-11*Z*,14*Z*-dienoate (**34**) and 14-BzA-8-*O*-docosahexaenoate (**37**). Compounds (**28–31** and **34**), which contain fatty acid side chains with 1-3 double bonds also showed low inhibitory activities with the exception of 14-BzA-8-*O*-eicosa-11*Z*-enoate (**33**), which contained a relatively long carbon side chain. Compounds with saturated acyl groups (**25–27, 32**) were inactive against COX-2 enzyme.

In the LTB$_4$ formation inhibitory assay all tested compounds showed activity to a certain extent. The highest activity was measured in case of 14-BzA-8-*O*-docosahexaenoate (**37**, 61.0% inhibition at 50 μM). All other LAs (**25–36**) expressed significant, but lower potency in 5-LOX mediated formation of LTB$_4$.

5.4.2 Assessment of the hERG-inhibiting ability of Aconitum alkaloids

The tested series of compounds represent a diversity of structural types, including bisnor- (C_{18}), nor- (C_{19}) and diterpene (C_{20}) alkaloids substituted with hydroxy, methoxy, keto, acetyl and various aromatic ester groups. All the compounds are *N*-ethyl-substituted with the exception of acovulparine (**57**), which contains an azomethine group. **Table 3** shows the measured hERG current inhibitions exerted by the compounds.

Table 3. Inhibition of hERG current by selected compounds at nominal concentration of 10 μM

Compound	Inhibition (%)	sem	*n*
acotoxicine (**1**)	17.3	3.3	5
aconitine (**5**)	44.9	7.4	5
isotalatizidine (**20**)	19.1	3.2	5
pyroaconitine (**24**)	18.7	1.6	4
10-hydroxy-8-*O*-methyltalatizamine (**38**)	15.8	1.1	8
hetisinone (**39**)	14.3	3.9	3
delcosine (**40**)	17.9	2.4	5
ajacine (**41**)	13.0	1.7	7
lycoctonine (**42**)	13.7	3.3	6
swatinine (**43**)	8.9	1.6	5
gigactonine (**44**)	38.0	7.4	5
14-BzA-8-*O*-palmitate (**48**)	39.6	5.6	7
aconosine (**49**)	15.3	3.5	6
dolaconine (**50**)	8.3	1.4	7
delectinine (**51**)	7.7	2.3	6
neolinine (**52**)	35.8	4.7	5
neoline (**53**)	14.4	3.7	5
acotoxinine (**54**)	6.5	2.2	6
songoramine (**55**)	36.4	5.4	4
songorine (**56**)	13.2	1.8	15
acovulparine (**57**)	10.8	2.3	8
septentriodine (**58**)	20.9	1.0	9
takaosamine (**59**)	12.5	2.9	6
14-desacetyl-18-demethylpubescenine (**60**)	6.5	1.9	6
haloperidol (at 1μM)	94.6	1.3	19

The highest hERG blockade activity was observed for the norditerpene diesters aconitine (**5**) and 14-BzA-8-*O*-palmitate (**48**), which exerted a 44.9 and a 39.6% inhibitory effect, respectively. The structurally similar pyroaconitine (**24**), with only one ester substituent, is a much less potent hERG channel inhibitor, with an 18.7% inhibitory effect. The further monoesters septentriodine (**58**), ajacine (**41**), acotoxinine (**54**) and dolaconine (**50**) also revealed only moderate K^+ channel potency, with 20.9, 13.0, 6.5 and 8.3% inhibition, respectively. Our results indicate that the substitution with two ester groups is an important structural feature for the K^+ channel activity. However, it can not be stated that the presence of an aryl ester group is required for the exertion of hERG activity, since ajacine (**41**) and acotoxinine (**54**) containing aromatic ester functionalities, do not possess high potency. As concerns the other non-esterified compounds based on the aconitane skeleton, gigactonine (**44**) and neolinine (**52**) exhibited marked hERG-inhibiting ability with 38.0 and 35.8% inhibitory activity, respectively. In the C_{20} alkaloid group, songoramine (**55**) (36.4% inhibition) was similarly as active as aconitine (**5**). Its close analogue, songorine (**56**), containing a 1-hydroxy group instead of 1,19-ether bridge, revealed only a low inhibitory potential on the hERG channel (13.2% inhibition).

5.4.3 Assessment of $Na_v1.2$ sodium channel activity of Aconitum alkaloids

This series of alkaloidal compounds represents diverse structural types as it is described in section *5.4.2*. **Table 4** shows the measured $Na_v1.2$ inhibitions exerted by the compounds.

Table 4. Inhibition of $Na_v1.2$ channels by selected compounds at nominal concentration of 10 µM

Compound	Inhibition (%)	sem	n
acotoxicine (**1**)	28	3.	8
aconitine (**5**)	25	3	8
isotalatizidine (**20**)	18	5	5
pyroaconitine (**24**)	57	2	7
10-hydroxy-8-*O*-methyltalatizamine (**38**)	16	7	8
hetisinone (**39**)	28	3	4
delcosine (**40**)	13	4	8
ajacine (**41**)	44	5	8
lycoctonine (**42**)	22	5	7
swatinine (**43**)	10	4	5
gigactonine (**44**)	2	4	7
aconosine (**49**)	18	6	5
dolaconine (**50**)	6	5	7
delectinine (**51**)	42	5	8
14-BzA-8-*O*-palmitate (**48**)	27	2	9
neolinine (**52**)	16	6	4
neoline (**53**)	15	4	7
acotoxinine (**54**)	14	4	4
songoramine (**55**)	15	3	8
songorine (**56**)	18	6	7
acovulparine (**57**)	30	4	6
septentriodine (**58**)	43	6.	7
takaosamine (**59**)	3	2	4
14-desacetyl-18-demethylpubescenine (**60**)	19	3	8

Pyroaconitine (**24**) exerted the highest inhibitory activity, for which the IC_{50} value was also determined (7.3±1.5 µM). Furthermore, ajacine (**41**), septentriodine (**58**), and delectinine (**51**) also demonstrated significant $Na_v1.2$ channel inhibition (44 – 42%) at 10 µM; several other compounds (acovulparine (**57**), acotoxicine (**1**), hetisinone (**39**), 14-BzA-8-*O*-palmitate (**48**), aconitine (**5**), and lycoctonine (**42**)) exerted moderate inhibitory activity (30 – 22%), while the rest of the tested alkaloids were considered to be inactive.

5.5 Determination of toxic alkaloid contents of processed *Radix aconiti*

The content of mesaconitine, aconitine, and hypaconitine has been determined in 16 commercial samples of processed aconite roots by HPLC and alkaloid titration (**Table 5**). These samples provide a near representative spectrum of processed TCM aconite drugs being available at Chinese and German markets.

Table 5. Origin of investigated aconite root samples and their alkaloid contents determined by HPLC analysis of toxic alkaloids (sum of mesaconitine, aconitine and hypaconitine) and by titration

Code	Simplified code for HPLC analyses	HPLC (%)	Titration (%)
Zhicaowu – Aconiti kusnezoffi praeparata			
37951	B3 (EDQM)	*not detectable*	*0.065*
37747	B4 (EDQM)	*0.007*	*0.013*
-	B6 (Hong Kong Baptist University)	*not detectable*	*0.194*
Zhichuanwu (Shanxi) – Aconiti praeparata (radix)			
32969	D1 (EDQM)	*0.054*	*0.207*
-	D2 (Hong Kong Baptist University)	*not detectable*	*0.120*
Aconiti *carmichaelii* radix praeparata/Aconiti radix praeparata			
28479	E1 (EDQM)	*0.013*	*0.062*
36890	E2 (EDQM)	*0.162*	*0.433*
Shanghai market – Aconiti radix praeparata/Aconiti radix lateralis praeparata			
37271	F2 – 2003	*not detectable*	*0.045*
37270	F3 – 2004	*not detectable*	*0.103*
37269	F4 - 2005	*not detectable*	*0.123*
37275	F5 – 2008	*0.125*	*0.342*
37272	F6 – 2009	*not detectable*	*0.070*
German market – Aconiti radix praeparata/Aconiti radix lateralis praeparata			
37276	G1 – 2004	*0.003*	*0.113*
37265	G2 – 2004	*0.026*	*0.129*
37263	G4 – 2007	*0.043*	*0.097*
37266	G5 – 2007	*0.011*	*0.116*

The HPLC method developed by us[82] provides good separation of the main toxic alkaloids (**Figure 5)** and therefore serves as basis for reliable quantitative analysis.

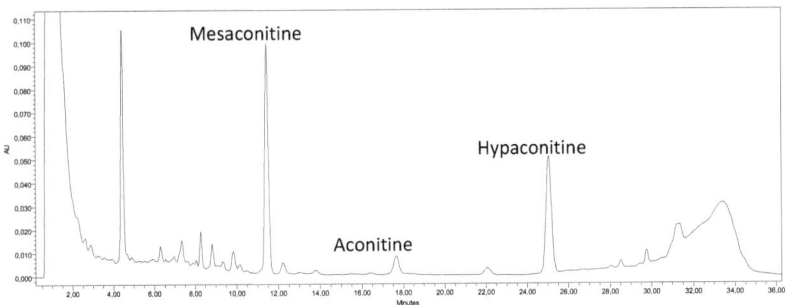

Figure 5. HPLC chromatogram of the extract prepared from a crude aconite root sample

For quantitative determination, linear regression analysis for aconitine was performed by the external standard method. The regression equation for aconitine was $y = 2\,999\,256\,695.1674x + 66\,930.4245$ (x stands for the amount of injected alkaloid in mg, y denotes the area under the curve). The correlation coefficient (R^2) was 0.9994. Using the HPLC method, in most of the samples no toxic alkaloids or only traces could be detected (**Table 5**). However, in four samples (D1, E2, F5, G4), >0.04% of aconitine, hypaconitine, and mesaconitine, the highest with a content of 0.162% (E2), were quantified. The alkaloid content determined by titration was considerably higher than determined by HPLC (**Figure 6**).

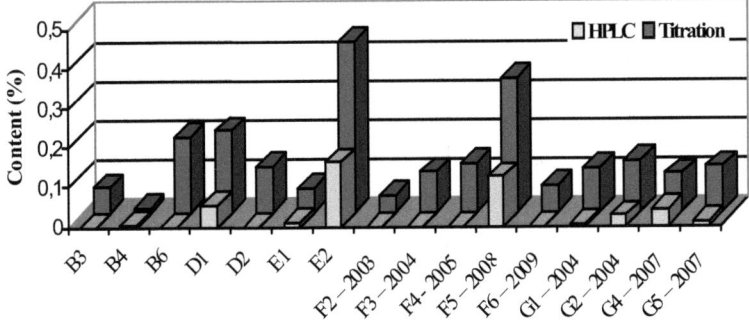

Figure 6. Comparison of content (%) of alkaloidal compounds measured by HPLC and titration.

Samples, in which toxic alkaloids were not detectable by HPLC, still contained up to 0.2% alkaloids according to titration. Careful processing obviously degrades the toxic alkaloids but does not remove all alkaloids. Total alkaloid titration determines not only the toxic diester alkaloids but also the monoester, unesterified, and LAs. There was no correlation between the low values (<0.02% HPLC), but there was a correlation in high values of alkaloid contents determined by the two methods (**Figure 7**). When taking into consideration only samples with measurable alkaloid content by HPLC, a correlation coefficient of 0.91 was found between the results of the two methods.

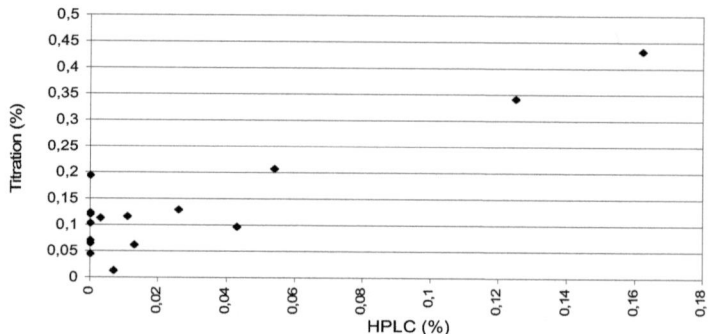

Figure 7. Correlation of the alkaloid contents by HPLC quantification and titration

6. DISCUSSION

DAs accumulated by *Aconitum* species have been in the focus of attention for several reasons. Amassed from classical and highly instrumentalised phytochemical studies the pieces of chemical information regarding the interesting, complex and variable structures of these compounds represent a body of information that by now literally surpassed the limits of encyclopaedic knowledge. Contrasting with this it should be noted that there is still no sufficient information on the pharmacology and toxicology of these alkaloids. These considerations and the fact that rare and/or endemic *Aconitum* species are partially or completely unexplored provided the primary motivation for the research group of HOHMANN *et al.* (Department of Pharmacognosy, University of Szeged) to start a research programme in 2001 dealing with *Aconitum* species native to the Carpatian Basin. In continuation of the work accomplished by CSUPOR *et al.* on *A. toxicum* and *A. vulparia,* and by HOHMANN and HAJDÚ *et al.* on *C. orientalis,* my work principally targeted the detailed phytochemical investigation of further two native monkshood species, *A. anthora.* and *A. moldavicum.*

Isolation of alkaloids from A. anthora *and* A. moldavicum

The phytochemical work was started with the screening of the alkaloid contents of the roots and above-ground parts of *A. anthora* and of the roots of *A. moldavicum.*

Taking into consideration the limited amount of the plant material, the above-ground parts and roots of *A. anthora* were examined together. In case of *A. moldavicum* the roots were processed. The plant materials were extracted with the amphipolar solvent, MeOH. In the case of *A. anthora,* a classical alkaloid isolation methodology based on solvent-solvent partitioning was applied. This method aims at the separation of alkaloids from neutral compounds. Further multiple-step chromatographic separation (including CC, VLC, GFC, PLC and CPC) on Al_2O_3 and Sephadex® LH-20 stationary phases resulted in the isolation of 3 pure compounds (**ANT–1-3**).

In the case of *A. moldavicum*, an isolation procedure in neutral medium was proposed to obtain the alkaloids. Reasons for the choice of this method were the facts that the extract of the roots did not contain chlorophyll; moreover, in neutral medium the risk of acidic or alkaline hydrolysis can be minimized. Initially, CC using an Al_2O_3 stationary phase was applied to remove polyphenolic compounds. After extensive chromatographic purification (including CC, VLC, GFC, PLC and CPC) with the use of Al_2O_3 and Sephadex® LH-20 and different solvent systems, 8 pure compounds were isolated (**AMO–1-8**).

Structure elucidation of isolated compounds

The structure determination of the isolated compounds was carried out by means of spectroscopic experiments. The most useful data regarding the structures originated from the 1D and 2D NMR (^{1}H-NMR, JMOD, ^{1}H-^{1}H COSY, HSQC, HMBC and NOESY) measurements. The already known compounds were identified by comparing their NMR data with those in the literature. The constitutions of the new compounds were elucidated by detailed analysis of the NMR spectra, supplemented with mass spectrometric experiments. The relative configurations were determined with the aid of NOESY experiments. As starting point, the stereochemistry of H-5 was considered to be β, as characteristic for DAs. Complete ^{1}H- and ^{13}C-NMR chemical shift assignments were made for the new compounds, and also to supplement or revise missing or incorrect data on the known compounds.

From *A. anthora*, 1 hetisane-type C_{20} [hetisinone (**39, ANT-3**)] and 2 aconitane-type C_{19} DAs [isotalatizidine (**20, ANT-1**) and 10-hydroxy-8-*O*-methyltalatizamine (**38, ANT-2**)] were identified (**Annex III**). The structure and relative configuration of the new alkaloid, 10-hydroxy-8-*O*-methyltalatizamine (**38, ANT-2**) were elucidated. ^{1}H- and ^{13}C-NMR chemical shift assignments were determined for this compound for the first time, with a corrected or supplemented assignment in the case of isotalatizidine (**20, ANT-1**). Hetisinone (**39, ANT-3**) was identified on the basis of the good agreement of measured and previously reported NMR data.

From *A. moldavicum*, 8 aconitane-type C_{19} DAs [delcosine (**40, AMO-1**), ajacine (**41, AMO-2**), lycoctonine (**42, AMO-3**), swatinine (**43, AMO-4**), gigactonine (**44, AMO-5**), cammaconine (**45, AMO-6**), and columbianine (**46, AMO7**)] were identified (**Annex III**). The complete structure and relative stereochemistry of the new alkaloid, 1-*O*-demethylswatinine (**47, AMO-11**) were determined. ^{1}H- and ^{13}C-NMR chemical shift assignments were determined for this compound for the first time; while for ajacine (**41, AMO-2**), and swatinine (**43, AMO-4**), complete ^{1}H chemical shift assignments were also carried out. All other alkaloids, namely delcosine (**40, AMO-1**), lycoctonine (**42, AMO-3**) gigactonine (**44, AMO-5**), cammaconine (**45, AMO-6**), and columbianine (**46, AMO-7**) were identified on the basis of the comparison of the measured and literature NMR data.

LAs: Semisynthesis and antiphlogistic activity testing

Processed aconite roots are widely used in Eastern medicinal systems, especially in TCM as painkillers and antirheumatic agents.[4] As long as aconitine-type DAs found in unprocessed roots are known to exhibit a broad spectrum of pharmacological activities, including antinociceptive and anti-inflammatory effects *in vitro*,[1,172] it is noteworthy that LAs are characteristic compounds of both processed and unprocessed aconite drugs. Their amount significantly increases in the course of the traditional processing of the drugs. Aconitine-type alkaloids (e.g. aconitine, hypaconitine, mesaconitine) are highly toxic, in contrast to LAs, which possess significantly less toxicity due to the presence of a long chain fatty acid moiety in the molecules at C-8.[31] In the last decade several analytical aspects of these compounds and processed aconite drugs[20,22,24,26,85,173] have been reported, but no detailed pharmacological studies were conducted in connection with their therapeutic relevance; and this phenomenon also means that less attention has been paid so far to the potential role of LAs in the pharmacological effects of processed aconite drugs.

A previous work carried out by CSUPOR *et al.* demonstrated that processing (usually boiling) of crude aconite roots decreases the amount of toxic alkaloids and increases the concentration of LAs. Therefore, toxic aconite alkaloids cannot be responsible for activity, but LAs may be.[31] The fact that the long chain fatty acid residues can reduce the high toxicity of DDAs, while still retaining the otherwise desirable antinociceptive and anti-inflammatory activities is of great importance. However, because the close structural similarities of compounds substituted with lipoid chains, these alkaloids have not been isolated to date in pure form from natural sources.

With the aim to evaluate the antiphlogistic potential of fatty acid substituted NDAs, a series of 13 aconitine derived LAs (**25–37**) as model substances were prepared semisynthetically, and subjected to *in vitro* anti-inflammatory assays, using the COX-1, COX-2 and LTB$_4$ formation inhibition models. The aim of this work was to gain information about the pharmacology of LAs and to conclude some structure-activity relationships concerning the nature of the esterifying fatty acids.[22]

In the reactions aconitine was transesterified by 13 different saturated and unsaturated fatty acids resulting the corresponding 14-BzA-8-O-esters and pyroaconitine. The COX-1, COX-2 and LTB$_4$ formation inhibitory activities of the compounds were investigated. In the COX-1 assay only compounds substituted with C$_{20}$ and C$_{22}$ fatty acid moiety (**32–37**) exhibited moderate activity, while all other LAs (**25–31**) were inactive on the COX-1 enzyme independently to the unsaturation and the length of the substituting fatty acid chain. In the COX-2 inhibition assay a correlation between the grade of unsaturation in the ester group and the enzyme inhibitory activity may be presumed, while regarding the 5-LOX assay a weak correlation could be noted between the activity and the length of the fatty acid chain (for graphical illustration of the activities of certain LAs see **Figure 8**).

These results demonstrating that LAs are not just by-products of a traditional drug processing method put further emphasis on analytical work, because without proper analytical mapping and quantitation neither the safety nor the efficacy of drugs of human use can be determined. It should be noted that currently no unambiguous pharmacological results are available studying concomitantly all the main alkaloid-type compounds of crude and processed aconite roots (DDAs, MAs, LAs and pyrolysates) in the same test system. Such work may be able to clearly distinguish the real roles of these compounds.

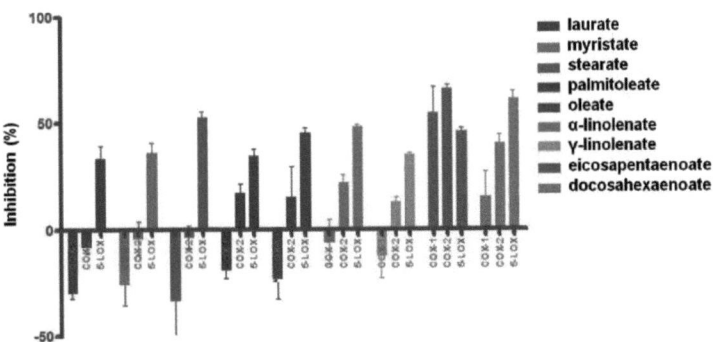

Figure 8. Anti-inflammatory activities of LAs

Investigation of ion channel inhibitory activities of selected compounds

Pharmacons that exhibit a use-dependent blockade of ion channels, have long been used, for example in the treatment of chronic neuropathic or inflammatory pain states, epilepsies, migraine and neurodegeneration related to ischaemia.[174,175] Because of the fact that in contrast with the exorbitant body of chemical information the common pharmacological knowledge on DAs can still be considered as rather poor, co-operative research programmes to gain further data on the pharmacological characteristics of DAs were initiated. Within the framework of this a selection of DA compounds was subjected to hERG and $Na_v1.2$ channel inhibitory assays.

Although most of the cardiovascular and neurologic effects and poisoning with *Aconitum* products are explained by the activation of Na^+ channels, the involvement of the K^+ channels in the pharmacological effects of DAs is not explored systematically. The human ether-à-go-go-related gene (hERG) encodes a cardiac voltage-gated K^+ channel ($K_v11.1$) that provides the major repolarising current (IKr) in phase 3 of the cardiac action potential. Most drugs that have been shown to increase the QT interval of the ECG at therapeutic doses incorporate IKr inhibition in their spectrum of effects. Prolongation of the QT interval can lead in rare cases to the polymorphic ventricular dysrhythmia called torsade de pointes (TdP) and sudden death.[176,177] Only one previous study has been reported the hERG channel inhibitory effect of the most commonly available DA, aconitine (5).[92]

Our data suggest that some structurally and functionally unrelated diterpene and NDAs may block the hERG K^+ channel, and may therefore act in cardiac action potential repolarisation, with prolongation of the QT interval, and increase of the risk of potentially fatal ventricular arrhythmias.[161] The toxic plasma concentration of aconitine (5) is 30-1000 times lower than its IC_{50} value on hERG channels, and its serious cardiac side effects are generally believed to be a result of Na^+ channel modulation.[178,179] However, for the other hERG-active alkaloids, the above difference in concentration is potentially small enough to exert their hERG effect during the therapeutic application of aconite drugs, which could result in adverse cardiovascular effects. The results obtained in this experiment do not allow establishment of a correlation between the structure and hERG channel activity of the alkaloids, but call the attention to the fact that relatively minor structural differences may cause considerable differences in efficacy[161]; and more importantly clearly indicate that the cardiovascular safety profile of *Aconitum* drugs should be evaluated taking into account the possible effect of DAs on hERG channels.

Developing subtype selective pharmacons that only block the channel isoforms specifically involved in the disease to be treated is a widely used strategy in modern drug research.[174] Theoretically, such compounds offer a relatively good side-effect profile, by leaving the channel subtypes involved in important physiological functions, such as the heart type $Na_v1.5$, untouched. Unfortunately, sodium channel blockers currently applied with different therapeutic indications discriminate poorly between $Na_v1.x$ subtypes and their tolerable therapeutic indices are rather the consequence of their use-dependent properties. Another important aspect is that compounds used therapeutically should devoid or have the least minimal activity against the hERG cardiac K^+ channel being involved in cardiac toxicity.[180,181]

Concerning the relationship between the structural features and the $Na_v1.2$ inhibiting activity of the studied alkaloids it seems relevant that the concomitant presence of a methoxy function at C-1, the presence of an oxygen-containing functionality at C-8, and the presence of a (substituted) aroyl function either at C-14 or at C-18 is necessary, with the parallel requirement that the number of oxygen functionalities in the molecule should be at least 4. Satisfaction of these stipulations can be seen in the case of all significantly (**41, 51, 58, 24**) and moderately active compounds (**42, 1, 57, 5, 24**). When C-1 methoxy is changed to hydroxyl, the inhibitory effects of the compounds decrease as it can be seen in the case of delectinine (**51**) (42% inhibition) and takaosamine (**59**) (3% inhibition). The only, nonetheless very interesting outlier compound is swatinine (**43**) (10% inhibition), which has a methoxy group at C-1; however, it is substituted with an unusual, extra hydroxyl at C-10. Without this latter substituent, the compounds substitution pattern is identical with acovulparine (**57**) (with the only structural difference occurring on the *N* atom) (30% inhibition). This surprising, and rather large activity difference allows to hypothesise the importance of hydroxy substitution at C-10 that may pose some kind of steric hindrance for the molecule when interacting with the channel receptor. However, it cannot be excluded that differences in the substitution of *N* atom may also be of importance regarding this difference in activity. Influence of the position of the aroyl functionality on the inhibitory activity is illustrated by the example of acotoxinine (**1**), which was found to be inactive, in spite of the fact that it has an aroyl functionality in the molecule, but it is positioned at C-8, not at C-14 or C-18, as it has been described above. These stipulations seem to be supported by a detailed QSAR analysis of 12 DAs, which revealed that the position of the aroyl/aroyloxy groups at C-4 or C-14 is the major determinant of the analgesic activity. Alkaloids with an aroyl or an aroyloxy group at C-14 exhibited an analgesic potency approximately 30 times higher than that of alkaloids with an aroyloxy group at C-4 in a model of acetic acid-induced writhing in rats.[182†] Previously it has also been proven that alkaloids with low affinity for the neurotoxin receptor site 2 of Na^+ channels lack antinociceptive action.[183†] As regards the three C_{20} DAs involved in this study only one (hetisinone, **39**) has been found to exert moderate activity, therefore no relevant deductions may be made concerning the structure-activity relationship of these compounds. It is noteworthy, though to mention that songorine (**56**) was found practically inactive, which is in line with the previous results of FRIESE *et al.* when observing that this compound has failed to modulate Na^+ channels.[87†] However, in their study no exact subtype specification was provided on the examined Na^+ channels. Further two aspects to be emphasised are that amongst the four most active compounds only one, delectinine (**51**) is not esterified (although satisfying all other presumably necessary conditions), and that both diester-type compounds (14-BzA-8-*O*-palmitate (**48**) and aconitine (**5**)) have been found to exert only moderate inhibitory activity on the studied particular channel subtype.

Considering our herein results, especially in light of our previously reported results on the hERG K^+ channel inhibitory activity of these DAs[161] it can be observed that certain compounds (14-BzA-8-*O*-palmitate (**48**), aconitine (**5**), gigactonine (**44**)) exerted noteworthy activity in both pharmacological tests, thus despite their promising $Na_v1.2$ inhibitory activity these compounds should be considered as potentially harmful by causing adverse cardiovascular effects through their hERG effect, nevertheless, some of these compounds show some selectivity over hERG channels. Their potential cardiovascular safety risk should be evaluated based on their hERG IC_{50} value and *in vivo* efficient free plasma concentration. Among the studied active alkaloids particularly those should be further pursued as model substances for pharmacological lead compounds as selective $Na_v1.2$ inhibitors, which exert minor or no hERG activity[161], i.e. ajacine (**41**) and delectinine (**51**).

Analysis of processed TCM aconite drugs for DA content

Traditional processing, which is a generally applied approach in the Far Eastern traditional medicinal systems, provides aconite drugs for human therapeutic use. Practically all analytical works on the topic demonstrated that cautious processing (usually boiling) of crude aconite roots decreases the amount of normal DAs and increases the concentration of LAs resulting in the reduction of toxicity of the drugs. Quality control criteria of pharmacopoeias are not always sufficient to warrant safety. In case of *Radix aconiti* and *Radix aconiti kusnezoffii*, the DA content is not limited in the PhPRC and the warning it stipulates ("be cautious about the unprocessed root taken orally") is not commensurable to the danger of toxicity. In the *Radix aconiti praeparata* and *Radix aconiti kusnezoffii praeparata* monographs, a colorimetric assay is used for the determination of DDAs (required content level should not be >0.15%), and a titrimetric assay for the determination of the total alkaloid content (required level should not be <0.20% of alkaloids, calculated as aconitine). The dosage of these two drugs is 1.5–3 g, which may contain as much as 4.5 mg DDAs.[84] Moreover, for *Radix aconiti lateralis praeparata*, the dose of which is 3–15 g, only the TLC analysis of aconitine is specified.[84] Taking into account that the minimum lethal dose of aconitine is 3–6 mg[184], it is obvious that only careful processing and quality control may warrant the safety of aconite containing medicinal products. Several homeopathic pharmacopoeias contain a monograph on *A. napellus*, and the essence of methods used for the qualification of its drugs, as demonstrated in the most prominent homeopathic pharmacopoeia, the *German Homeopathic Pharmacopoeia*, is the titrimetric determination of the total alkaloid content.[185] Since the titrimetric determination provides no direct information on the toxic DDA content of aconite roots, more reliable methods are required for the quality control of processed plant material. Therefore the aim of our work was to develop a quick and simple HPLC method for the quality control of aconite roots with comparable or better reliability to those of the previously published methods[186-189], to compare the results of titration method with HPLC analysis of the toxic alkaloids.[82]

In most of the commercial samples, toxic alkaloids were undetectable, or only traces were found by the applied HPLC method. However, the fact that in four samples (D1, E2, F5, G4), toxic aconite alkaloid levels could be detected above 0.04%, highlights the consideration that these alkaloid contents are high enough to question the safety of the samples concerned. Samples with mesaconitine, aconitine, and hypaconitine content below the HPLC detection limit still contained up to 0.2% alkaloids determined by titration. With this comparison of results of HPLC and titrimetric analyses no correlation was found between the two methods.[82] The sample preparation and HPLC analysis method developed by us offers a quick and reliable possibility to determine the quantity of the most important toxic alkaloids of *A. carmichaelii* and *A. kusnezoffii* and was developed with the aim of providing a proper analytical tool for pharmacopoeial aconite drug analysis.

7. SUMMARY

Within the frame of the research programme of the Department of Pharmacognosy the present doctoral thesis targeted the thorough phytochemical examination of samples (collected from Hungarian populations) of *A. anthora* and *A. moldavicum* resulting in the isolation and structure elucidation of 2 new, and 9 known DAs. One new compound, 10-hydroxy-8-*O*-methyltalatizamine (**ANT-2, 38**), together with the known compounds, isotalatizidine (**ANT-1, 20**) and hetisinone (**ANT-3, 39**) were obtained from *A. anthora*; while another new alkaloid, 1-*O*-demethylswatinine (**AMO-8, 47**), together with seven known compounds, delcosine (**AMO-1, 40**), ajacine (**AMO-2, 41**), lycoctonine (**AMO-3, 42**), swatinine (**AMO-4, 43**), gigactonine (**AMO-5, 44**), cammaconine (**AMO-6, 45**), and columbianine (**AMO-7, 46**) were isolated from *A. moldavicum*. With the exception of isotalatizidine (**ANT-1, 20**) all alkaloids were reported from the investigated taxa for the first time.

The structures were established by means of HRESIMS, 1D and 2D NMR spectroscopy, including ^{1}H-^{1}H COSY, NOESY, HSQC and HMBC experiments, resulting in complete and unambiguous ^{1}H and ^{13}C assignments for all isolated compounds, and the reassignment, supplementation or revision of missing or incorrect data on the known compounds, isotalatizidine (**ANT-1, 20**), ajacine (**AMO-2, 41**), and swatinine (**AMO-4, 43**).

To ensure the availability of LAs for pharmacological studies, a formerly published semisynthetic approach using aconitine and fatty acids as synthons of natural origin have been applied in a refined form. As a result, a series of 13 aconitine derived LAs (**25–37**) were prepared semisynthetically, purified by using a variety of chromatographic methods, and in the frame of co-operation, subjected to *in vitro* anti-inflammatory assays, using the COX-1, COX-2 and LTB$_4$ formation inhibition models. Results obtained from these assays reinforced the presumption that LAs may play a role in the anti-inflammatory effect of processed aconite roots. The production and antiphlogistic testing of these LAs were reported for the first time and certain structure-activity relationships concerning the nature of esterifying fatty acids on the anti-inflammatory activity has also been proposed for the first time.

Within the framework of co-operations, twenty-four C$_{19}$ and C$_{20}$ DAs representing the structural diversity (different skeletal types and substitution patterns) of *Aconitum* alkaloids (including several of those obtained from the herein discussed 2 taxa) were subjected to measure their hERG and Na$_v$1.2 channel inhibitory effects by whole-cell patch clamp technique. Although the results obtained in the hERG-inhibiting ability assay do not allow establishment of a correlation between the structure and hERG channel activity of the alkaloids, but clearly indicate that the cardiovascular safety profile of *Aconitum* drugs should be evaluated taking into account the possible effect of DAs on hERG K$^+$ channels.

On the contrary, our results obtained in the Na$_v$1.2 channel inhibitory activity assay made possible to propose certain structure-activity relationships regarding the structural features that may influence the channel activity of DAs. These observations may open new perspectives in the research of channel subtype specific compounds, especially when considering that DAs comprise large and chemically diverse and biologically remarkably active compound repertory. The herein reported specific inhibitory activity of certain DAs on Na$_v$1.2 channel is intriguing especially in light of the facts that in different geographical regions of the world some DA containing *Delphinium* and *Aconitum* species have long been used as both unprocessed and processed drugs of folk and traditional medicine for the same analgetic and antiepileptic indications[4,190] as those Na$_v$1.2 inhibitory drug candidate compounds.[174]

In conclusion, it can be stated that distinctly minor substitutional changes seem to have rather firm impact on the inhibitory activity of DAs both on hERG and Na$_v$1.2 channels. It should be noted that experimental pharmacological results obtained in relation to the inhibitory activities of DAs on both the hERG K$^+$ channel and the specific Na$^+$ channel subtype, Na$_v$1.2, were reported for the first time.

As consequence of the harmonization of TCM and modern medicinal systems aconite preparations are also attracting increasing interest of today's Western medicine, which is strikingly illustrated by the fact that despite the potentially hazardous nature of these drugs their incorporation into the Ph. Eur. is currently in progress. In order to contribute to the improvement of quality control of these drugs, we have developed a simple, quick and reliable HPLC method to measure the toxic DA in aconite drugs and validated the method by quantifying 16 commercial samples of processed *A. carmichaelii* and *A. kusnezofii* drugs. The results obtained with the HPLC assessment were compared with a titrimetric analysis of the total alkaloid contents and the conclusion has been drawn that our HPLC method may be considered as proper tool for purity and quality test in pharmacopoeial monographs.

To sum up all these phytochemical and pharmacological results from a pharmacognostic point of view it should be emphasised that DAs expressing notable bioactivities should be regarded as valuable lead compounds worthy of further research and development; and with regard to natural sources all plant species being able to "economically" produce these exceptional substances should be attributed distinct importance and attention to ensure protection and detailed scientific investigation.

Novel scientific results – résumé:

1. Phytochemical investigation of *A. anthora* and *A. moldavicum* resulted in the isolation of 2 new [1-*O*-demethylswatinine (**11, AMO-8**) and 10-hydroxy-8-*O*-methyltalatizamine (**2, ANT-2**)], and 9 known DAs [hetisinone (**3, ANT-3**), isotalatizidine (**1, ANT-1**), delcosine (**4, AMO-1**), ajacine (**5, AMO-2**), lycoctonine (**6, AMO-3**), swatinine (**7, AMO-4**), gigactonine (**8, AMO-5**), cammaconine (**9, AMO-6**) and columbianine (**10, AMO-7**)], which – with the exception of one compound – was isolated from these species for the first time; the latter taxon was investigated chemically for the first time;

2. Through extensive 1D and 2D NMR studies structure elucidation and complete and unambiguous ^{1}H and ^{13}C assignments was carried out for all isolated

compounds; reassignment, supplementation or revision of missing or incorrect data was performed for the already known compounds;

3. 13 Aconitine-derived LAs were prepared semisynthetically and purified chromatographically by a methodology developed by us;

4. Anti-inflammatory assays with the semisynthetic LAs were carried out on the COX-1, COX-2 and LTB_4 formation inhibition models to establish for the first time certain antiphlogistic structure-activity relationships concerning the effect of the nature of the esterifying fatty acid moiety;

5. Pharmacological bioassays on the hERG and $Na_v1.2$ channels were carried out by using a selected series of DAs to establish for the first time certain structure-activity relationships concerning the effect of the tested DAs on these channels;

6. As a result of these pharmacological assays ajacine (**AMO-2**) and delectinine, which are selective $Na_v1.2$ inhibitors, and at the same time exert minor or no hERG activity were identified as perspective compounds for further pharmacological analyses and as potential lead compounds.

7. Determination of the toxic DA content in 16 authentic processed *Radix aconiti* samples was performed by a HPLC method developed by us, 4 potentially toxic commercial samples were identified; comparison of our results with the widely applied titrimetry revealed the superiority of our method in terms of accuracy and therapeutic relevance.

8. REFERENCES

1   Ameri A *Prog Neurobiol* **1998**; *56*:211-223
2   Fan ZC, Zhang ZQ *J Chem Crystallogr* **2008**; *38*:895-899
3   Csupor D *Investigation of the diterpene alkaloids of Aconitum species native to the Carpathian Basin*, In: *Annals of Albert Szent-Györgyi Medical & Pharmaceutical Center* (ed. Fülöp F), Vol. *143*, JATEPress, Szeged, **2007.**
4   Singhuber J, Zhu M, Prinz S, Kopp B *J Ethnopharmacol* **2009**; *126*:18-30
5   Borhidi A *A zárvatermők fejlődéstörténeti rendszertana*, Nemzeti Tankönyvkiadó, Budapest, **1995**. p 169
6   Jabbour F, Renner S *Mol Phylogenet Evol* **2012**; *62*:928-942
7   Utelli AB, Roy BA, Baltisberger M *Plant Syst Evol* **2000**; *224*:195-212
8   Tutin TG, Burges NA, Chater AO, Edmondson JR, Heywood VH, Moore DM, Valentine DH, Walters SM, Webb DA, eds., *Flora Europaea* Vol. *1.*, Cambridge University Press, Cambridge, **1993**. pp 254-256
9   Farkas G, ed., *Magyarország védett növényei*, Mezőgazda Kiadó, Budapest, **1999**. pp 106-107
10 Soó R *A magyar flóra és vegetáció rendszertani, növényföldrajzi kézikönyve II.*, Akadémiai Kiadó, Budapest, **1966**. p 41
11 Liu H, Katz A *J Nat Prod* **1996**; *59*:135-138
12 Csupor D, Forgo P, Wenzig EM, Bauer R, Hohmann J *J Nat Prod* **2008**; *71*:1779-1782
13 Kolak U, Öztürk M, Özgökçe F, Ulubelen A *Phytochemistry* **2006**; *67*:2170-2175
14 Dalton DR *The Alkaloids: The Fundamental Chemistry - A Biogenetic Approach*, in *Studies in Organic Chemistry* (ed. Gassman PG), Vol. *7*, Marcel Dekker, New York, **1979.**
15 Southon LW, Buckingham J *Dictionary of Alkaloids*, Chapman and Hall, London, New York, **1989.**
16 Wang FP, Liang XT C_{20}-diterpenoid alkaloids In: The Alkaloids: Chemistry and Biology (ed. Cordell GA), Vol. *59*, , Academic Press, **2002**. pp 61-65
17 Kitagawa I, Yoshikawa M, Chen ZL, Kobayashi K *Chem Pharm Bull* **1982**; *30*:758-761
18 Wang FP, Chen QH, Liu XY *Nat Prod Rep* **2010**; *27*:529-570
19 Borcsa B, Dezső Cs, Forgo P, Widowitz U, Bauer R, Hohmann J *Nat Prod Commun* **2011**; *6*:527-536
20 Xu Q, Wang Y, Liu C, Liu Z, Liu S *Anal Sci* **2003**; *19*:1599-1603

21 Hanuman JB, Katz A *J Nat Prod* **1994**; *57*:105-115

22 Sun W, Song F, Cui M, Liu S *Planta Med* **1999**; *65*:432-436

23 Sun WX, Liu SY, Liu ZQ, Song, FR Fang SP *Rapid Commun Mass Spectrom* **1998**; *12*:821-824

24 Shim SH, Lee SY, Kim JS, Son KH, Kang SS *Arch Pharm Res* **2005**; *28*:1239-1243

25 Shim SH, Kim JS, Kang SS *Chem Pharm Bull* **2003**; *51*:999-1002

26 Yue H, Pi Z, Song F, Liu Z, Cai Z, Liu S *Talanta* **2009**; *77*:1800-1807

27 Wang Y, Liu ZQ, Song FR, Liu SY *Rapid Commun Mass Spectrom* **2002**; *16*:2075-2082

28 Wang Y, Song FR, Xu QX, Liu ZQ, Liu SY *J Mass Spectrom* **2003**; *38*:962-970

29 Wang J, van der Heijden R, Spijksma G, Reijmers T, Wang M, Xu G, Hankemeier T, van der Greef J *J Chromatogr A* **2009**; *1216*:2169-2178

30 Wu W, Liang ZT, Zhao ZZ, Z.W. Cai ZW *J Mass Spectrom* **2007**; *42*:58-69

31 Csupor D, Wenzig EM, Zupkó I, Wölkart K, Hohmann J, Bauer R *J Chromatogr A* **2009**; *1216*:2079-2086

32 Schmitz G, Ecker J *Prog Lipid Res* **2008**; *47*:147-155

33 Wang Y, Shi L, Song FR, Liu ZQ, Liu SY *Rapid Commun Mass Spectrom* **2003**; *17*:279-284

34 Zhang S, Zhao G, Lin G *Phytochemistry* **1999**; *51*:333-336

35 Díaz JG, Ruiz JG, Herz W *Phytochemistry* **2004**; *65*:2123-2127

36 Xiong L, Peng C, Xie XF, Guo L, He CJ, Geng Z, Wan F, Dai O, Zhou QM *Molecules* **2012**; *17*:9939-9946

37 Shim SH, Kim JS, Son KH, Bae KH, Kang SS *J Nat Prod* **2006**; *69*:400-402

38 Suzgec S, Bitis L, Sozer U, Ozcelik H, Zapp J, Kiemer AK, Mericli F, Mericli AH *Chem Nat Compd* **2009**; *45*:287-289

39 Xu Y, Guo ZJ, Wu N *Fitoterapia* **2010**; *81*:1091-1093

40 Meriçli AH, Meriçli F, Ulubelen A, Bahar M, Ilarslan R, Algül G, Desai HK, Teng Q, Pelletier SW *Pharmazie* **2000**; *55*:696-698

41 Pirildar S, Unsal Gurer C, Kocyigit M, Zapp J, Kiemer AK, Meriçli AH *Chem Nat Compd* **2013**; *48*:1115-1116

42 Mariani C, Braca A, Vitalini S, De Tommasi N, Visioli F, Fico G *Phytochemistry* **2008**; *69*:1120-1126

43 Wangchuk Ph, Keller PA, Pyne SG, Taweechotipatr M, Tonsomboon A, Rattanajak R, Kamchonwongpaisan S *J Ethnopharmacol* **2011**; *137*:730-742

44 Wangchuk P, Bremmer JB, Samten, Skelton BW, White AH, Rattanajak R, Kamchonwongpaisan S *J Ethnopharmacol* **2010**; *130*:559-562

45 Wangchuk P, Bremmer JB, Samosorn S *J Nat Prod* **2007**; *70*:1808-1811

46 Zafar S, Ahmad MA, Siddiqui TA *Fitoterapia* **2002**; *73*:535-556

47 Wada K, Nihira M, Ohno Y *J Ethnopharmacol* **2006**; *105*:89-94

48 Ono T, Hayashida M, Uekusa K, Lai CF, Hayakawa H, Nihira M, Ohno Y *Leg Med* **2009**; *11*:132-135

49 Burdel'naya EV, Zhunusova MA, Turmukhambetov AZ, Seidakhmetova RB, Shul'ts EE, Gatilov YV, Adekenov SM *Chem Nat Compd* **2012**; *47*:1032-1034

50 Chodoeva A, Bosc JJ, Guillon J, Decendit A, Petraud M, Absalon Ch, Vitry Ch, Jarry Ch, Robert J *Bioorg Med Chem* **2005**; *13*:6493-6501

51 Song MJ, Kim H *J Ethnopharmacol* **2011**; *137*:167-175

52 Jeong HJ, Whang WK, Kim IH *Planta Med* **1997***; 63:*329-334

53 Batbayar N, Enkhzaya S, Tunsag J, Batsuren D, Rycroft DS, Sproll S, Bracher F *Phytochemistry* **2003**; *62*:543-550

54 Shrestha PM, Katz A *J Nat Prod* **2000**; *63*:2-5

55 Shrestha PM, Katz A *J Nat Prod* **2004**; *67*:1574-1576

56 Meriçli AH, Yazici S, Eroglu-Ozkan E, Sen B, Kurtoglu S, Ozcelik H, Zapp J, Kiemer AK, Mericli F *Chem Nat Compd* **2012**; *48*: 525-526

57 Meriçli F, Meriçli AH, Ulubelen A, Desai HK, Pelletier SW *J Nat Prod* **2001**; *64*:787-789

58 Meriçli F, Meriçli AH, Tan N, Özçelik H, Ulubelen A *Sci Pharm* **1999**; *67*:313-318

59 Meriçli AH, Meriçli F, Desai HK, Ilarslan R, Ulubelen A, Pelletier SW *Pharmazie* **2001**; *55*:696-698

60 Xiong J, Tan NH, Ji ChJ, LuY, Gong NB *Tetrahedron Lett* **2008**; *49*:4851-4853

61 Jiang SH, Yang PM, Zhou H, Zhu DY *Planta Med* **2002**; *68*:1147-1149

62 Tang QF, Ye WC, Liu JH, Y ChH *Phytochem Lett* **2012**; *5*:397-400

63 Jiang SH, Wang HQ, Li YM, Lin SJ, Tan JJ, Zhu DY *Chin Chem Lett* **2007**; *18*:409-411

64 Shen Y, Zuo AX, Jiang ZY, Zhang XM, Wang HL, Chen JJ *Helv Chim Acta* **2011**; *94*:268-272

65 Zhang F, Peng SL, Liao X, Yu KB, Ding LS *Planta Med* **2005**; *71*:1073-1076

66 Sun JY, Li TC *J Chem Res* **2009**; *5*:306-307

67 Zhou XL, Chen DL, Chen QH, Wang FP *J Nat Prod* **2005**; *68*:1076-1079

[68] Hong YH, Peacher WG *Chinese Herb Medicine and Therapy*, Oriental Healing Institute of U.S.A., Los Angeles, **1986**. p 162

[69] Bensky D, Gamble A *Chinese Herbal Medicine*, Eastland Press, Seattle, **1991**. pp 428-430

[70] Gao LM, Wei XM, Yang L *Chin Chem Lett* **2005**; *16*:475-478

[71] Cao JX, Li LB, Jiang SP, Tian RR, Chen XL, Peng SL, Zhang J, Zhu HJ *Helv Chim Acta* **2008**; *91*:1954-1960

[72] Yang XD, Yang S, Yang J, Zhao JF, Zhang HB, Li L *Helv Chim Acta* **2008**; *91*:569-573

[73] He YQ, Ma ZY, Wei XM, Du BZh, Jin ZX, Yao BH, Gao LM *Fitoterapia* **2010**; *81*:929-931

[74] He YQ, Yao BH, Ma ZhY *J Pharm Anal* **2011**; *1*:57-59

[75] Wang YB, Huang R, Zhang HB, Li L *Helv Chim Acta* **2005**; *88*:1081-1084

[76] Qu SJ, Tan CH, Liu ZL, Jiang SH, Yu L, Zhu DY *Phytochem Lett* **2011**; *4*:144-146

[77] Yang S, Yang XD, Zhao JF, Zhang HB, Li L *Helv Chim Acta* **2007**; *90*:1160-1164

[78] Kang Y, Łuczaj ŁJ, Ye S *Genet Resour Crop Evol* **2012**; *59*:1569-1575

[79] *Pharmacopoea Hungarica*, Pesti Könyvnyomda-Részvény-Társulat, Budapest, **1871**.

[80] *Pharmacopoea Hungarica Editio Secunda*, Pesti Könyvnyomda-Részvény-Társulat, Budapest, **1888**.

[81] *Pharmacopoea Hungarica Editio Tertia*, Pesti Könyvnyomda-Részvény-Társulat, Budapest, **1909**.

[82] Csupor D, Borcsa B, Heydel B, Hohmann J, Zupkó I, Ma Y, Widowitz U, Bauer R *Pharm Biol* **2011**; *49*:1097-1101

[83] EDQM homepage:
https://extranet.edqm.eu/4DLink1/4DCGI/Web_View/mono/2470;
https://extranet.edqm.eu/4DLink1/4DCGI/Web_View/mono/2429 Accessed: 17 March **2013**

[84] *Pharmacopoeia of the People's Republic of China.* Vol. *I*, People's Medical Publishing House, Beijing, China, **2005.**

[85] Wang J, van der Heijden R, Spijksma G, Reijmers T, Wang M, Xu G, Hankemeier T, van der Greef J *J Chromatogr A* **2009**; *1216*:2169-2178

[86] Hanuman JB, Katz A *J Nat Prod* **1994**; *57*:105-115

[87] Friese J, Gleitz J, Gutser UT, Heubach JF, Matthiesen T, Wilffert B, Selve N *Eur J Pharmacol* **1997**; *337*:165-174

88 Voss LJ, Voss JM, McLeay L, Sleigh JW *Eur J Pharmacol* **2008**; *584*:291-296
89 Hardick DJ, Blagbrough IS, Cooper G, Potter BVL, Critchley T, Wonnacott S *J Med Chem* **1996**; *39*:4860-4866
90 Sultankhodzhaev MN, Khan MTH, Moin M, Choudhary MI, Atta-Ur-Rahman *Nat Prod Res* **2005**; *19*:517-522
91 Shaheen F, Ahmad M, Khan MTH, Jalil S, Ejaz A, Sultankhodjaev MN, Arfan M, Choudhary MI, Atta-ur-Rahman *Phytochemistry* **2005**; *66*:935-940
92 Guo L, Dong Z, Guthrie H *J Pharmacol Toxicol Methods* **2009**; *60*:130-151
93 Desai HK, Hart BP, Caldwell RW, Jiangzhong-Huang, Pelletier SW *J Nat Prod* **1998**; *61*:743-748
94 Mazur NA, Ivanova LA, Pavlova TS, *Biull Vsesoiuznogo Kardiol Nauchn Tsentra AMN SSSR* **1986**; *9*:30-33
95 Liu SS, Liu IM, Lai MC, Cheng JT *J Ethnopharmacol* **2005**; *99*:379-383
96 Wang J, van der Heijden R, Spruit S, Hankermeier T, Chan K, van der Greef J, Xu G, Wang M *J Ethnopharmacol* **2009**; *126*:31-41
97 Wang DP, Lou HY, Huang L, Hao XJ, Liang GY, Yang ZC, Pan WD *Bioorg Med Chem Lett* **2012**; *22*:4444-4446
98 Shu H, Arita H, Hayashida M, Sekiyama H, Hanaoka K *J Ethnopharmacol* **2006**; *103*:398-405
99 Shu H, Arita H, Hayashida M, Chiba S, Sekiyama H, Hanaoka K *J Ethnopharmacol* **2006**; *106*:263-271
100 Shu H, Hayashida M, Chiba S, Sekiyama H, Kitamura T, Yamada Y, Hanaoka K, Arita H *J Ethnopharmacol* **2007**; *113*:560-563
101 Shu H, Hayashida M, Huang W, An K, Chiba S, Hanaoka K, Arita H *J Ethnopharmacol* **2008**; 117:158-165
102 Wu G, Huang W, Zhang H, Li Q, Zhou J, Shu H *J Ethnopharmacol* **2011**; *136*:254-259
103 Wang CF, Gerner P, Wang SY, Wang GK *Anesthesiology* **2007**; *107*:82-90
104 Kawata Y, Ma CM, Meselhy MR, Nakamura N, Wang H, Hattori M, Namba T, Satoh K, Kuraishi Y *J Trad Med* **1999**; *16*:15-23
105 Kukel CF, Jennings KR *Can J Physiol Pharmacol* **1994**; *72*:104-107
106 Hardick DJ, Cooper G, Scott-Ward T, Blagbrough IS, Potter BV, Wonnacott S *FEBS Lett* **1995**; *365*:79-82
107 Wada K, Ishizuki S, Mori T, Bando H, Murayama M, Kawahara N *Biol Pharm Bull* **1997**; *20*:978-982

[108] Atta-ur-Rahman, Nashreen A, Akhtar F, Shekhani MS, Clardy J, Parvez M, Choudhary MI *J Nat Prod* **1997**; *60*:472-474

[109] González-Coloma A, Guadaño A, Gutiérrez C, Cabrera R, de la Peña E, Reina M *J Agric Food Chem* **1998**; *46*:286-290

[110] Ulubelen A, Meriçli AH, Meriçli F, Neşet K, Ferizli AG, Emekci M, Pelletier SW *Phytoter Res* **2001**; *15*:170-171

[111] González-Coloma A, Reina M, Guadaño A, Martínez-Díaz R, Díaz JG, García-Rodriguez J, Alva A, Grandez M *Chem Biodivers* **2004**; 1327-1335

[112] Şener B, Orhan I, Özçelik B *ARKIVOC* **2007**; *VII*:265-272

[113] Chodoeva A, Bosc JJ, Guillon J, Costet P, Decendit A, Mérillon JM, Léger JM, Jarry C, Robert J *Eur J Med Chem* **2012**; *54*:343-351

[114] Wada K, Hazawa M, Takahashi K, Mori T, Kawahara N, Kashiwakura *J Nat Prod* **2007**; *70*:1854-1858

[115] Wada K, Ohkoshi E, Morris-Natschke, Bastow KF, Lee KH *Bioorg Med Chem Lett* **2012**; *22*:249-252

[116] Gao F, Li YY, Wang D, Huang X, Liu Q *Molecules* **2012**, *17*:5187-5194

[117] He YQ, Ma ZY, Wei XM, Liu DJ, Du BZ, Yao BH, Gao LM *Chem Biodivers* **2011**; *8*:2104-2109

[118] Gao LM, Yan HY, He YQ, Wei XM *J Integr Plant Biol* **2006**; *48*:364-369

[119] Chang JG, Shih PP, Chang CP, Chang JY, Wang FY, Tseng J *Planta Med* **1994**; *60*:576-578

[120] Li M, He J, Jiang LL, Ng ESK, Wang H, Lam FFY, Zhang YM, Tan NH, Shaw PC *J Ethnopharmacol* **2013**; *147*:122-127

[121] Kolak U, Öztürk M, Özgökçe F, Ulubelen A *Phytochemistry* **2006**; *67*:2170-2175

[122] Fu M, Wu M, Qiao Y, Wang Z *Pharmazie* **2006**; *61*:735-741

[123] Chan TYK *Forensic Sci Int* **2012**; *223*:25-27

[124] Ito K, Tanaka S, Funayama M, Mizugaki M *J Anal Toxicol* **2000**; *24*:348-353

[125] Bonnici K, Stanworth D, Simmonds MSJ, Mukherjee E, Ferner RE *Lancet* **2010**; *376*:1616

[126] Lin CC, Chan TY, Deng JF *Ann Emerg Med* **2004**; *43*:574-579

[127] Kolev ST, Leman P, Kite GC, Stevenson PC, Shaw D, Murray VS *Human Exp Toxicol* **1996**; *15*:839-842

[128] Tai YT, But PPH, Young K, Lau CP *Lancet* **1992**; *340*:1254-1256

[129] Chan TYK, Tomlinson B, Tse LKK, Chan JCN, Chan WWM, Critchley JAJH *Vet Hum Toxicol* **1994**; *36*:452-455

[130] Zeng ZP, Jiang JG *Brit J Pharmacol* **2010**; *159*:1374-1391

[131] Kolev ST, Leman P, Kite GC, Stevenson PC, Shaw D, Murray VS *Hum Exp Toxicol* **1996**; *15*:839-842

[132] Chan TYK *Clin Toxicol* **2009**; *47*:279-285

[133] Zhao Z, Liang Z, Chan K, Lu G, Lee ELM, Chen H, Li L *Planta Med* **2010**; *76:*1975-1986

[134] Shaw D *Planta Med* **2010**; *76:*20012-2018

[135] Lai CK, Poon WT, Chan YW *J Anal Toxicol* **2006**; *30*:426-433

[136] Chan TYK *Hum Exp Toxicol* **2011**; *30*:2023-2026

[137] Munnecom THC, van Kraaij DJW, van Westreenen JC *Int J Cardiol* **2011**: *152*:e37–e39

[138] Chan TYK *Forensic Sci Int* **2012**; *222*:1-3

[139] Arlt EM, Keller T, Wittmann H, Monticelli F *Leg Med* **2012**; *14*:154-156

[140] Larabi K, Soulillou A, Mattys M, Lucchini MJ, Fanton Y, Tafani B *Presse Med* **2013**; *42*:353-354

[141] Liu Q, Zhuo L, Liu L, Zhu S, Sunnassee A, Liang M, Zhou L, Liu Y *Forensic Sci Int* **2011**; *212*:e5–e9

[142] Dominois-Heraud AM, Schmitt C, De Matteis O, Tichadou L, de Haro L *Ann Fr Anesth Reanim* **2012**; *31*:262-270

[143] Lu GH Dong ZQ, Wang Q, Qian GS, Huang WH, Jiang ZH, Leung KSY, Zhao ZZ *Planta Med* **2010**; *76*:825-830

[144] Ito K, Ohyama Y, Konishi Y, Tanaka S, Mizugaki M *Planta Med* **1997**; *63*:75-79

[145] Kaneko R, Hattori S, Furuta S, Hamajima M, Hirata Y, Watanabe K, Seno H, Ishii A *J Mass Spectrom* **2006**; *41*:810-814

[146] Wang ZH, Guo D, He Y, Hu C, Zhang J *Phytochem Anal* **2004**; *15*:16-20

[147] Ohta H, Seto Y, Tsunoda N *J Chromatography* B **1997**; *691*:351-356

[148] Hayashida M, Hayakawa H, Wada K, Yamada T, Nihira M, Ohno Y *Leg Med* **2003**; *5*:S101-S104

[149] Zhang HG, Sun Y, Duan MY, Chen YJ, Zhong DF, Zhang HQ *Toxicon* **2005**; *46*:500-506

[150] Wang Z, Wang Z, Wen J, He Y *J Pharm Biomed Anal* **2007**; *45*:145-148

[151] Nakae H, Fujita Y, Igarashi T, Tajimi K, Endo S *Biomed Res* **2008**; *29*:225-231

[152] Usui K, Hayashizaki Y, Hashiyada M, Nakano A, Funayama M *Leg Med* **2012**; *14*:126-133

[153] Chung KKW, Chen SPL, Ng SW, Mak TWL, Leung KSY *Talanta* **2012**; *97*:491-498

[154] Tang L, Gong Y, Lv C, Ye L, Liu L, Liu Z *J Ethnopharmacol* **2012**; *141*:736-741

[155] Tang L, Ye L, Lv C, Zheng Z, Gong Y, Liu Z *Toxicol Lett* **2011**; *202*:47-54

[156] Ye Ling, Tang L, Gong Y, Lv C, Zheng Zh, Jiang Z, Liu Z *Xenobiotica* **2011**; *41*:46-58

[157] Xin Y, Pi Z, Song F, Liu Z, Liu S *Chin J Chem* **2012**, *30*:656-664

[158] Fan YF, Xie Y, Liu L, Ho HM, Wong YF, Liu ZQ, Zhou H *J Ethnopharmacol* **2012**; *141*:701-708

[159] Bai Y, Desai HK, Pelletier SW *J Nat Prod* **1994**; *57*:963-970

[160] Borcsa B, Widowitz U, Csupor D, Forgo P, Bauer R, Hohmann J *Fitoterapia* **2011**; *82*:365-368

[161] Forgo P, Botond B, Csupor D, Fodor L, Berkecz R, Molnár VA, Hohmann J *Planta Med* **2011**; *77*:368-373

[162] van Beek TA, van Veldhuizen A, Lelyveld GP, Piron I, Lankhorst PP *Phytochem Anal* **1993**; *4*:261-268

[163] Boido V, Edwards OE, Handa KL, Kolt RJ, Purushothaman KK *Can J Chem* **1984**; *62*:778-784

[164] Pelletier SW *Alkaloids: Chemical and Biological Perspectives* Vol. *II.*, Wiley & Sons, New York, Chichester, Brisbane, Toronto, Singapore, **1984**. pp 399

[165] De la Fuente G, Ruíz-Mesia L *Phytochemistry* **1995**; *39*:1459-1465

[166] Gonzalez AG, de la Fuente G, Reina M, Díaz R, Timón I *Phytochemistry* **1986**; *25*:1971-1973

[167] Hajdú Z, Forgó P, Löffler B, Hohmann J *Biochem Sys Ecol* **2005**; *33*:1081-1085

[168] Csupor D, Forgo P, Máthé I, Hohmann J *Helv Chim Acta* **2004**; *87*:2125-2130

[169] Pelletier SW *Alkaloids: Chemical and Biological Perspectives* Vol. *II.*, Wiley & Sons, New York, Chichester, Brisbane, Toronto, Singapore, **1984**. pp 205-462

[170] Hanuman JB, Katz A *Phytochemistry* **1994**; *36*:1527-1535

[171] De la Fuente G, Orribo T, Gavin JA, Acosta RD *Heterocycles* **1993**; *36*:1455-1458

[172] González-Coloma A, Reina M, Mediaveitia A, Guadaño A, Santana O, Martínez-Díaz R, Ruis-Mesía L, Alva A, Grandez M, Díaz R, Gavín JA, de la Fuente G *J Chem Ecol* **2004**, *30*:1393-1408

[173] Yang M, Sun J, Lu Z, Chen G, Guan S, Liu X, Jiang B, Ye M, Guo DA *J Chromatogr A* **2009**; *1216*:2045-2062

[174] Tarnawa I, Bölcskei H, Kocsis P *Recent Pat CNS Drug Disc* **2007**; *2*:57-78

[175] Lai J, Hunter JC, Porreca F *Curr Opin Neurobiol* **2003**; *13*:291-297

[176] De Bruin ML, Pettersson M, Meyboom RHB, Hoes AW, Leufkens HGM *Eur Heart J* **2005**; *26*:590-597

[177] Sanguinetti MC, Tristani-Firouzi M *Nature* **2006**; *440*:463-469

[178] Beike J, Frommherz L, Wood M, Brinkmann B, Köhler H *Int J Legal Med* **2004**; *118*:289-293

[179] Ohta H, Seto Y, Tsunoda N *J Chromatogr B* **1997**; *691*:351-356.

[180] Liang J, Brochu RM, Cohen CJ, Dick IE, Felix JP, Fisher MH, Garcia ML, Kaczorowski GJ, Lyons KA, Meinke PT, Priest BT, Schmalhofer WA, Smith MM, Tarpley JW, Williams BS, Martin WJ, Parsons WH *Bioorg Med Chem Lett* **2005**; *15*:2943-2947

[181] Baker MD, Wood JN *Trends Pharmac Sci* **2001**; *22*:27-31

[182] Bello-Ramírez AM, Buendía-Orozco J, Nava-Ocampo AA *Fundam Clin Pharmacol* **2003**; *17*:575-80

[183] Gutser UT, Friese J, Heubach JF, Matthiesen T, Selve N, Wilffert B, Gleitz J *Naunyn Schmiedeberg's Arch Pharmacol* **1998**; *357*:39-48

[184] Caravati EM, McCowan CL, Marshall SW *Plants*. In: Dart RC, ed. *Medical Toxicology*. Lippincott Williams & Wilkins, Philadelphia, **2004**. p 1701

[185] *German Homeopathic Pharmacopoeia*. Vol. *1*, Medpharm, Scientific Publishers Stuttgart, Germany, **2005**.

[186] Jiang ZH, Xie Y, Zhou H, Wang JR, Liu ZQ, Wong YF, Cai X, Xu HX, Liu L *Phytochem Anal* **2005**; *16*:415-421

[187] Xie Y, Jiang ZH, Zhou H, Xu HX, Liu L *J Chromatogr A* **2005**; *1093*:195-203

[188] Wang Z, Wen J, Xing J, He Y *J Pharm Biomed Anal* **2006**; *40*:1031-1034

[189] Kang XQ, Fan ZC, Zhang ZQ *J Chromatogr Sci* **2010**; *48*:860-865

[190] *Medicinal Plants in China*. WHO Regional Publications Western Pacific Series 2, WHO Regional Office for the Western Pacific, Manila, **1989**. p 5

ACKNOWLEDGEMENTS

I wish to express my most grateful thanks to my chief supervisor, *Prof. Dr. Judit Hohmann*, for introducing me to the world of diterpene alkaloids and phytochemical research. I am greatly indebted to her for her professional guidance, continuous support and uncompromised help she has provided for me.

I wish to express my humble and most sincere gratitude to my co-supervisor, *Dr. Dezső Csupor* assistant professor for his never ending encouragement, warm hearted reassurance, uncountable advices and profound humanity, with which he not only reinforced and inspired me, but also strengthened my perseverance at certain times of extreme difficulties.

I wish to thank to *Prof. Dr. Imre Máthé* and *Prof. Dr. Judit Hohmann*, former and current heads of the Department of Pharmacognosy, for placing the facilities at my disposal and providing me with all the opportunities to carry out my experimental work.

I owe special thanks to *Dr. Attila Molnár V.* (Department of Botany, University of Debrecen) for the identification and collection of the plant material.

My special thanks are due to my co-authors, namely *Dr. Péter Forgó* for the NMR measurements; *Ute Widowitz*, *Barbara Heydel* and *Prof. Dr. Rudolf Bauer* (Karl-Franzenz University, Graz, Austria) for the anti-inflammatory tests and titrimetric measurements, *Dr. László Fodor* (Gedeon Richter Plc.) for the $Na_v1.2$ and hERG tests, and *Róbert Berkecz* (Department of Medical Chemistry, University of Szeged) for the mass spectra.

I am very grateful to all staff members at the Department of Pharmacognosy for their help and assistance, in particular to *Erzsébet Hadárné Berta* laboratory assistant, *Dr. Andrea Vasas* and *Dr. Dóra Rédei* assistant professors, and *Dr. Katalin Veres* research fellow for their unselfish help and advices given to my experimental work.

Last but not least I would love to express my grateful thanks to my most beloved mother, family and friends for all their help, support and encouragement, without which I could not accomplish this work.

Annex I. Types of diterpene alkaloids

C₁₈-diterpene alkaloids

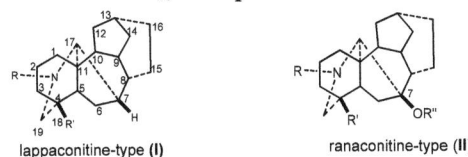

lappaconitine-type (I)   ranaconitine-type (II)

C₁₉-diterpene alkaloids

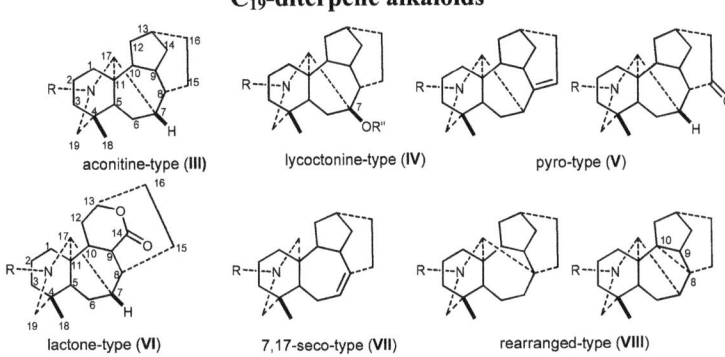

aconitine-type (III)   lycoctonine-type (IV)   pyro-type (V)

lactone-type (VI)   7,17-seco-type (VII)   rearranged-type (VIII)

C₂₀-diterpene alkaloids

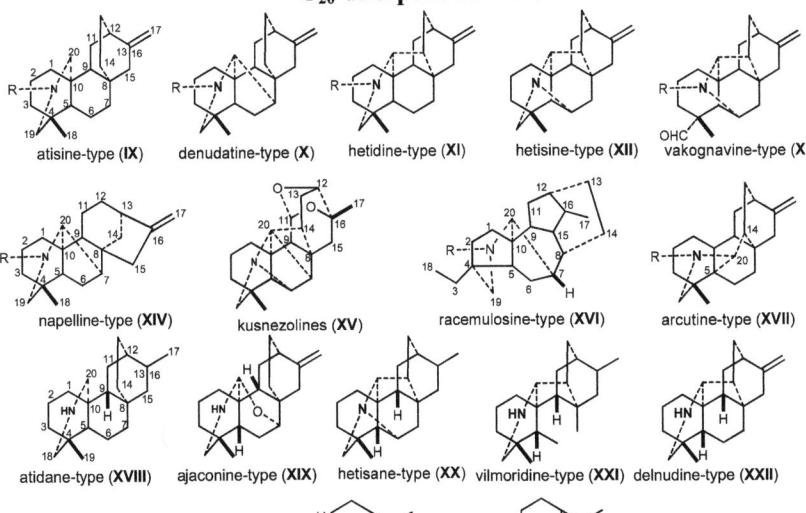

atisine-type (IX)   denudatine-type (X)   hetidine-type (XI)   hetisine-type (XII)   vakognavine-type (XIII)

napelline-type (XIV)   kusnezolines (XV)   racemulosine-type (XVI)   arcutine-type (XVII)

atidane-type (XVIII)   ajaconine-type (XIX)   hetisane-type (XX)   vilmoridine-type (XXI)   delnudine-type (XXII)

veatchine-type (XXIII)   14,20-cycloveatchine-type (XXIV)

Annex II. Previously isolated or semisynthesised alkaloids tested in the bioassays of hERG and Na$_v$1.2 channels

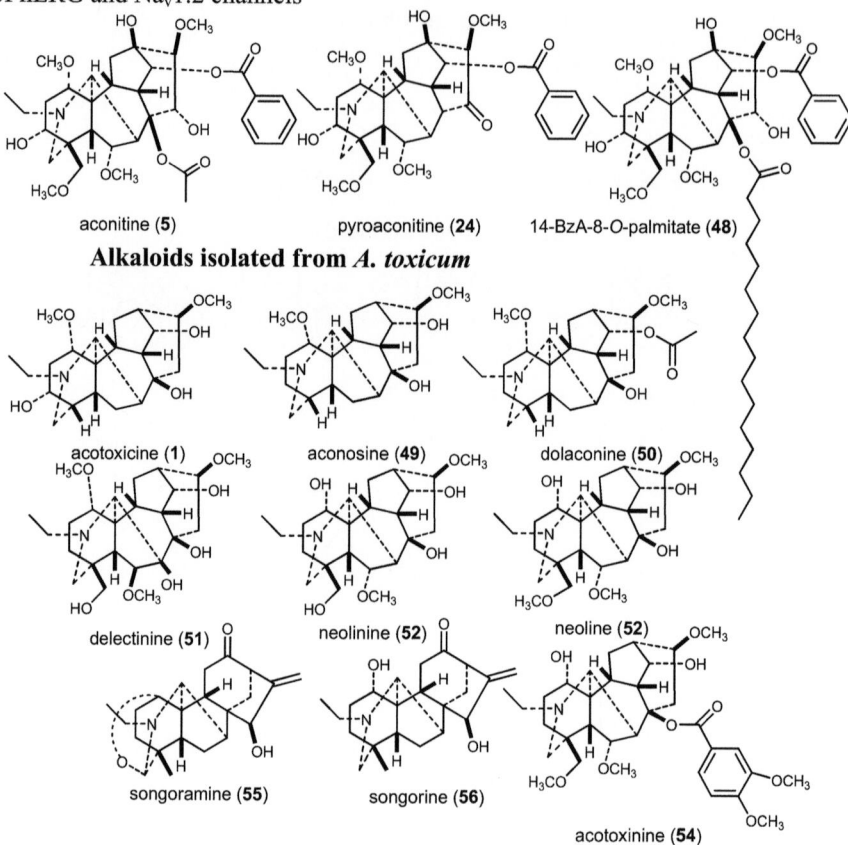

aconitine (**5**)

pyroaconitine (**24**)

14-BzA-8-O-palmitate (**48**)

Alkaloids isolated from *A. toxicum*

acotoxicine (**1**)

aconosine (**49**)

dolaconine (**50**)

delectinine (**51**)

neolinine (**52**)

neoline (**52**)

songoramine (**55**)

songorine (**56**)

acotoxinine (**54**)

Alkaloids isolated from *A. vulparia*

Alkaloids isolated from *Consolida orientalis*

acovulparine (**57**)

septentriodine (**58**)

gigactonine (**44**)

delcosine (**40**)

takaosamine (**59**)

14-desacetyl-18-demethyl-pubescenine (**62**)

Annex III.

Alkaloids isolated from *A. anthora*

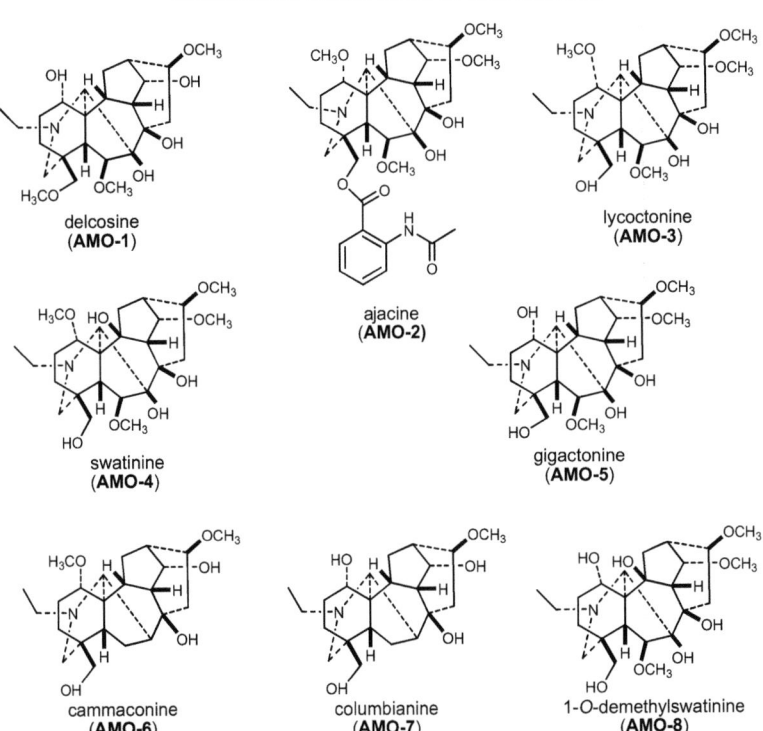

isotalatizidine
(**ANT-1**)

10-hydroxy-8-O-methyltalatizamine
(**ANT-2**)

hetisinone
(**ANT-3**)

Alkaloids isolated from *A. moldavicum*

delcosine
(**AMO-1**)

ajacine
(**AMO-2**)

lycoctonine
(**AMO-3**)

swatinine
(**AMO-4**)

gigactonine
(**AMO-5**)

cammaconine
(**AMO-6**)

columbianine
(**AMO-7**)

1-O-demethylswatinine
(**AMO-8**)

Annex IV – **Table A.** NMR spectral data of 10-hydroxy-8-O- methyltalatizamine (**ANT**-2) isolated from *A. anthora* [500 MHz (^{1}H), 125 MHz (^{13}C), CDCl$_3$, δ (ppm) (J = Hz)]

Atom	^{1}H	^{13}C	HMBC (H→C)	NOESY (H No.)
1	3.75 t (4.8)	78.6	2, 10, 11, 17	2β, 3β, 5, 1-OCH$_3$,
2α	2.28 ddd (4.8, 10.2, 19.5)	26.0	1	3α, 19a
2β	2.04 m			1
3α	1.75 m	32.4	4, 5	2α, 18b, 19a
3β	1.42 ddd (5.0, 10.8, 15.0)		4, 5	1, 5
4	-	38.4	-	-
5	1.84 s	42.1	4, 17	1, 3β, 9, 18a,
6β	1.82 dd (18.8, 7.7)	24.1	8	9
6α	1.47 m		8, 11	18a, 18b, 19a, 19b
7	2.41 d (7.4)	39.4	5, 8, 9, 11	6α, 15α, 17, 19b, 8-OCH$_3$
8	-	77.0	-	-
9	1.99 d (4.7)	55.5	8, 10, 14	5, 6β, 14
10	-	81.1	-	-
11	-	54.8	-	-
12α	2.66 d (16.1)	39.1	10, 11, 13, 14	16, 17
12β	1.71 dd (16.1, 8.2)		10, 14, 16	13, 14
13	2.50 m	38.6	9, 10, 15	12β, 14, 16-OCH$_3$
14	4.59 t (4.5)	73.3		9, 12β, 13
15α	2.19 dd (15.7, 9.0)	35.3	8, 9, 13, 16	7, 16, 17
15β	2.10 dd (15.7, 4.5)		8, 13, 16	8-OCH$_3$
16	3.33 dd (8.9, 4.5)	81.8		12α, 15α, 17
17	2.87 s	63.3	5, 6, 10, 11, 19, 20	7, 12α, 15α, 16, 19a, 21
18a	3.13 d (9.0)	79.5	3, 4, 5, 19	5, 6α, 19b
18b	3.01 d (9.0)		3, 4, 5, 19	3α, 6α, 19a, 19b
19a	2.54 d (11.5)	52.9	3, 4, 5, 17	2α, 3α, 6α, 17, 18b, 21
19b	2.02 d (11.6)		3, 4, 18, 20	6α, 18a, 18b
20a	2.50 m	49.4	17, 19, 21	21
20b	2.41 dq (19.4, 7.4)		17, 19, 21	21
21	1.08 t (7.4)	13.6	20	17, 19a, 20a, 20b
1-OMe	3.28 s	56.0	1	1
8-OMe	3.14 s	48.2	8	7, 15β
16-OMe	3.37 s	56.4	16	13
18-OMe	3.31 s	59.5	18	18a, 18b

Annex IV – **Table B.** NMR spectral data of 1-*O*-demethylswatinine (**AMO-8**) isolated from *A. moldavicum* [500 MHz (^{1}H), 125 MHz (^{13}C), CDCl$_3$, δ(ppm) (J = Hz)]

Atom	^{1}H	^{13}C	HMBC (C→H)	NOESY
1	4.06 brs	69.8	3α	2α, 3α, 3β, 12α
2α	1.72 m	34.6		1
2β	1.68 m			5, 9
3α	1.98 m	26.7	18a, 18b, 19a, 19b	1
3β	1.47 m			1
4	-	38.2	6, 5, 18a, 19a, 19b, 3α, 2β	-
5	2.15 s	40.8	6, 17, 18b, 19a, 19b	2β, 6, 9
6	4.03 s	91.0	5, 6-OMe, 17	5, 18a, 19a
7	-	87.2	5, 6, 15α, 15β	-
8	-	76.8	14, 6, 8-OH, 9, 17, 15α, 15β	-
9	2.83 d (2.5)	53.7	15α , 12α	2β, 5, 14, 8-OH
10	-	83.0	9, 12α	-
11	-	54.1	5, 6, 17, 12α	-
12α	2.34 d (15.1)	41.0	16	1, 16, 17
12β	1.95 dd (15.1, 7.5)			13, 14
13	2.58 dd (7.4, 4.6)	38.2	9, 12α, 15α	12b, 14
14	4.11 t (4.6)	82.9	14-OMe, 12α	9, 12β, 13
15α	2.63 dd (14.7, 8.5)	34.6	8-OH, 9	16, 17
15β	1.80 dd (14.7, 8.3)			
16	3.26 dd (8.5, 8.3)	82.0	9, 12α, 12β , 15α, 15β, 16-OMe	12α, 15α, 17
17	2.74 d (1.7)	66.4	5, 19a, 19b	12α, 15α, 16, 20a, 21
18a	3.70 d (10.5)	66.9	3α, 19a, 19b	6
18b	3.40 d (10.5)			
19a	2.49 d (11.5)	57.1	5, 17, 20	6
19b	2.45 d (11.5)			
20a	2.97 dq (19.4, 7.3)	50.3	19a	17
20b	2.85 dq (19.4, 7.3)			
21	1.11 t (7.3)	14.0	20a, 20b	17
1-OMe	-	-	-	-
6-OMe	3.43 s	58.0	6	8-OH
14-OMe	3.45 s	57.8	14	
16-OMe	3.35 s	56.3	16	8-OH
1-OH	7.50 s	-	-	
8-OH	3.99 s	-	-	9, 6-OMe, 16-OMe

Annex IV – Table C. NMR spectral data of isotalatizidine (**ANT-1**) isolated from *A. anthora* [500 MHz (^{1}H), 125 MHz (^{13}C), CDCl$_3$, δ (ppm) (J = Hz)]

Atom	^{1}H	^{13}C
1	3.72 s	72.3
2α	1.58 m (2H)	29.8
2β		
3α	1.86 m	26.9
3β	1.65 m	
4	-	37.3
5	1.82 m	41.7
6β	1.90 m	24.9
6α	1.65 m	
7	2.09 m	45.2
8	-	74.1
9	2.22 t (6.0, 5.7)	46.7
10	1.84 m	44.1
11	-	48.6
12α	1.62 m	28.3
12β	2.04 m	
13	2.35 m	39.9
14	4.22 t (5.0)	75.9
15α	2.42 m	42.3
15β	2.07 m	
16	3.38 dd (9.1, 4.6)	81.9
17	2.82 s	64.0
18a	3.14 d (8.8)	79.1
18b	3.00 d (8.8)	
19a	2.34 d (10.9)	56.6
19b	2.10 d (10.9)	
20a	2.55 dq (12.7, 7.1)	48.5
20b	2.45 dq (12.7, 7.1)	
21	1.12 t (7.1)	13.0
1-OMe	-	-
8-OMe	-	-
16-OMe	3.34 s	56.3
18-OMe	3.31 s	59.4

Annex IV – Table D. ^{1}H NMR spectral data of ajacine (**AMO-2**) and swatinine (**AMO-4**) isolated from *A. moldavicum* [500 MHz (^{1}H), CDCl$_3$, δ (ppm) (J = Hz)]

Atom	AMO-4	AMO-7
1	3.58 t (7.9, 9.2)	3.74 brs
2α	2.15 m	1.67 m
2β	2.15 m	1.58 m
3α	1.60 m	2.10 m
3β	1.50 m	1.91 m
4	-	-
5	1.95 s	1.84 m
6	3.89 s	1.92 m
		1.62 m
7	-	2.10 m
8	-	-
9	2.88 d (4.7)	2.23 t (5.9)
10	-	1.87 m
11	-	-
12α	3.07 d (15.7)	2.06 m
12β	1.70 m	1.60 m
13	2.49 dd (7.8, 4.6)	2.34 m
14	4.10 t (4.8, 4.5)	4.24 t (5.0)
15α	2.64 dd (15.3, 10.0)	2.42 dd (16.3, 9.3)
15β	1.70 m	2.07 m
16	3.17 t (8.5, 7.2)	3.39 dd (9.2, 4.7)
17	2.84 d (2.1)	2.82 s
18a	3.64 d (10.7)	3.46 d (10.5)
18b	3.37 d (10.9)	3.29 d (10.5)
19a	2.60 d (12.0)	2.36 d (10.7)
19b	2.32 d (12.0)	2.08 d (10.8)
20a	2.90 m	2.55 dq (12.4, 7.1)
20b	2.81 dq (19.2, 7.2)	2.46 dq (12.4, 7.1)
21	1.05 t (7.2)	1.13 t (7.1)
1-OMe	3.26 s	-
6-OMe	3.45 s	-
14-OMe	3.44 s	-
16-OMe	3.33 s	3.35 s
1-OH	-	3.03 brs*
8-OH	4.08 s	

*Not assigned signal.

Appendix I

The thesis is based on the following publications:

I.   Forgo P, **Borcsa B**, Csupor D, Fodor L, Berkecz R, Molnár VA, Hohmann J
Diterpene alkaloids from *Aconitum anthora* and assessment of the hERG-
inhibiting ability of *Aconitum* alkaloids
Planta Med **2011**; *77*:368-373

II.   **Borcsa B**, Widowitz U, Csupor D, Forgo P, Bauer R, Hohmann J
Semisynthesis and pharmacological investigation of lipo-alkaloids prepared
from aconitine
Fitoterapia **2011**; *82*:365-368

III.   **Borcsa B**, Csupor D, Forgo P, Widowitz U, Bauer R, Hohmann J
Aconitum lipo-alkaloids – semisynthetic products of the traditional medicine
Nat Prod Commun **2011**; *6:*527-536

IV.   Csupor D, **Borcsa B**, Heydel B, Hohmann J, Zupkó I, Ma Y, Widowitz U,
Bauer R
Comparison of a specific HPLC determination of toxic aconite alkaloids in
processed Radix aconiti with a titration method of total alkaloids
Pharm Biol **2011**; *49*:1097-1101

APPENDIX II
LIST OF PUBLICATIONS RELATED TO THE THESIS

I.   Forgo P, **Borcsa B**, Csupor D, Fodor L, Berkecz R, Molnár VA, Hohmann J
Diterpene alkaloids from *Aconitum anthora* and assessment of the hERG-
inhibiting ability of *Aconitum* alkaloids
Planta Med **2011**; *77*:368-373

II.   **Borcsa B**, Widowitz U, Csupor D, Forgo P, Bauer R, Hohmann J
Semisynthesis and pharmacological investigation of lipo-alkaloids prepared
from aconitine
Fitoterapia **2011**; *82*:365-368

III.   **Borcsa B**, Csupor D, Forgo P, Widowitz U, Bauer R, Hohmann J
Aconitum lipo-alkaloids – semisynthetic products of the traditional medicine
Nat Prod Commun **2011**; *6:*527-536

IV.   Csupor D, **Borcsa B**, Heydel B, Hohmann J, Zupkó I, Ma Y, Widowitz U,
Bauer R
Comparison of a specific HPLC determination of toxic aconite alkaloids in
processed Radix aconiti with a titration method of total alkaloids
Pharm Biol **2011**; *49*:1097-1101

V.   **Borcsa B**, Fodor L, Csupor D, Forgo P, Molnár VA, Hohmann J
Diterpene alkaloids from *Aconitum moldavicum* and assessment of $Na_v1.2$
sodium channel activity of *Aconitum* alkaloids
Planta Med **2014**; *80*:231-236